Patterns of Development

Patterns of Development

Resources, Policy and Economic Growth

Richard M. Auty

Edward Arnold
A member of the Hodder Headline Group
LONDON NEW YORK MELBOURNE AUCKLAND

First published in Great Britain 1995 by
Edward Arnold, a division of Hodder Headline PLC,
338 Euston Road, London NW1 3BH

British Library Cataloguing in Publication Data
A catalogue record for this book is available from the British Library

ISBN 0 340 59502 7 (Pb)
ISBN 0 340 62511 2 (Hb)

Typeset in 10/12 pt Century Old Style by Scribe Design, Gillingham, Kent.
Printed and bound in Great Britain by Mackays of Chatham PLC, Chatham, Kent

Contents

List of tables

List of figures

Preface

This book is aimed at undergraduates and grew out of a second year course at Lancaster University. Its interdisciplinary focus should appeal to students of geography, economics and politics. It adopts a strongly empirical approach but draws on a range of theories, mostly economic, which are presented in a way that should be readily accessible to the layman.

The book has three key objectives. The first is to provide a classification of developing countries which permits general statements to be made without drifting into the sweeping over-generalization which often characterizes discussion about development issues. Five types of developing country are identified, based upon country size and natural resource endowment: low-income African, low-income large Asian, mid-income Latin American, the mineral economy and newly industrializing East Asian.

The second objective is to take stock of the numerous development experiments to date. The efficacy of these experiments was sternly tested by the upheavals in the international economy associated with the 1973 and 1979 oil shocks. A consensus has since emerged on which policies are most likely to yield rapid and equitable economic growth. This book argues that the new consensus relegates many previous policies to history and opens up a new agenda for debate. The new debate includes such issues as environmental sustainability and also democracy.

The third main objective of the book is to make a start in incorporating sustainable development into the new policy consensus. The focus remains largely at the national level and deals with flow resources (like soils and forests) and fund resources (ores and fossil fuels), as well as the 'residual sinks' (the use of land, water and air to absorb pollution). The emerging debate between advocates of strong and weak sustainability is outlined in the final chapter of the book.

The book is basically structured around a comparison of the postwar economic development of the five main types of developing country. Their diverging economic performance is explained largely by reference to internal policies, but particular attention is paid to the impact of a country's natural resource endowment on policy choice. The tendency for well-endowed countries to under-perform in relation to their potential is noted and explained in terms of the 'resource curse' thesis. The policy discussion focuses on the choice of the broad development strategy, rather than associated social policies. Successful development makes it possible to tackle a wide range of social and environmental problems, whereas failure subverts such goals to the needs of short-term economic crisis management.

Within this overall framework, the book pursues five central themes: rural neglect, income inequality, hyper-urbanization, unequal terms of trade and the role of government. Each theme is discussed with particular reference to one of the five types of country, noting the implications for the other four country types. The book is also structured so that a second subset of themes can be pursued. Rural issues are the concern of Chapters 2–4, whereas industrial/urban issues are the focus in Chapters 5–7 and international trade – notably the varied responses of developing countries to the postwar economic order – comes to the fore in Chapters 8–11.

The five basic themes are explored as follows: first, rural neglect is analysed with reference to low-income Africa, and especially that region's arid and semi-arid areas. Second, the related theme of unequal income distribution is tackled with reference to low-income Asia, which has also suffered from urban bias. The third theme, hyper-urbanization and urban primacy, is analysed with reference to the mid-income countries of Latin America, the most urbanized set of countries. International trade, the fourth theme, is examined with reference both to the influential Singer/Prebisch thesis concerning the unequal terms of trade, and also the more radical dependency theory critique of the multinational corporation. Finally, the liberal international economic order is explored with reference to the success of the East Asian countries. The Asian development 'escalator' is analysed, tracing the diffusion of a widening range of manufacturing from Japan to the four Asian 'Dragons' – Hong Kong, Singapore, Taiwan and South Korea – and thence to Malaysia, Thailand and Indonesia.

The book concludes on an optimistic note: rapid and equitable growth can be achieved and environmental degradation minimized, provided political systems can be devised which mediate between competing domestic interest groups. But such political solutions must be tailored to the specific conditions of individual countries. Unlike economic and environmental issues, political solutions are too complex to be readily amenable to 'general' answers. Hence this book provides a stepping stone to a more detailed analysis of economic development which is more appropriately undertaken at a regional or national level.

R. Auty 1994

Acknowledgements

Marlene Mcnaught typed the tables and Nicky Shadbolt drew the illustrations. Both exhibited great dedication and skill in the face of the author's tight deadlines and loose handwriting. At Edward Arnold, Laura McKelvie provided constant encouragement, without which the book would not have materialized. Finally, my wife cheerfully helped in proofreading and in a thousand other ways, as is her wont.

List of abbreviations

CACM	Central American Common Market
EC	European Community
ECLA	Economic Commission for Latin America
EDP	Environmental domestic product
FAO	Food and Agricultural Organization
GDP	Gross domestic product
GNP	Gross national product
GWP	Gross world product
HCI	Heavy and chemical industry
IADB	Inter-American Development Bank
ICOR	Incremental capital output ratio
IMF	International Monetary Fund
ISI	Import substitution industrialization
LIEO	Liberal international economic order
MNC	Multinational corporation
MSF	Mineral stabilization fund
NDP	Net domestic product
NIC	Newly industrializing country
NIEO	New international economic order
OECD	Organization for Economic Cooperation and Development
OPEC	Organization of Petroleum Exporting Countries
RBI	Resource-based industrialization
SOE	State-owned enterprise
WHO	World Health Organization
WRI	World Resources Institute

PART 1

Overview: worlds within the Third World

1

Diverging development: types of developing country

Diverging interpretations of postwar development

Sharp differences of opinion exist concerning the postwar progress of global economic development. A basic distinction can be made between orthodox and structuralist perspectives. Orthodox observers, like the economic historian Rostow (1990), or the neo-liberal economist Lal (1983), tend to be optimistic: they see economic growth diffusing throughout the global economy via the 'trickle-down effects' of trade. The trickle-down effects are the economic stimuli created in the developing countries by the industrial countries' demand for raw materials and agricultural commodities.

Those on the structuralist wing of the debate tend to be far less optimistic about the process of economic development. They point to a widening gap between rich and poor countries. Within the structuralist perspective, a useful distinction may be made between radicals and reformers (Chilcote 1974). The radicals include dependency theorists, such as Gundar Frank, who believe that growth in developing countries is dependent on international capital. The reformers include economists closer to the mainstream, such as Singer and Prebisch.

Both the structuralist groups agree that the richer countries use their political power to set the rules governing international trade, boosting their own trading interests at the expense of the poorer countries. But whereas the radicals prescribe revolutionary solutions, the reformers look to intervention by developing country governments to temporarily restrain market forces while they build up competitive industrial sectors behind high external tariffs.

Examples of orthodox and structuralist thinking will clarify the major differences between them. A very influential paradigm developed within the orthodox perspective is Rostow's five-stage model of economic growth. On the structuralist side, the core/periphery model of the global economy has been widely deployed. Both models give useful insights into the development process, but they also provide a salutary warning of the dangers of over-generalizing about economic development.

Rostow's five-stage growth model

Rostow (1990) proposed a five-stage model of economic growth in the mid-1950s (Fig. 1.1), in a book subtitled *A Non-Communist Manifesto*. Although his model has been extensively criticized it nevertheless provides useful insights into changes in the role of economic sectors and technology as development proceeds. Rostow argues that countries can expect to go through five stages of economic growth, beginning with the traditional economy and progressing through the stages of pre-conditioning, economic take-off and drive to maturity, before reaching the fifth and final stage of maturity.

The principal characteristics of the traditional economy are first, an agricultural economic base which employs 70–80 per cent of the population in low-productivity farming with little surplus if any for sale. A second feature of the traditional economy is a highly feudalistic social order in which change is unusual and technology is static. The basic unit of spatial organization is the village, yielding a landscape of repetitive, self-contained economic cells. The individual's spatial horizons are limited, as portrayed for example, for late-nineteenth century Ibo society in Chinua Achebe's (1958) novel, *Things Fall Apart*.

The second (pre-conditioning) stage of development sees the emergence of a strong central government, able to establish law and order and to guarantee the safety of merchants to trade. In consequence, a group in society emerges which is able to recognize and exploit the new opportunities for more efficient production. A third feature of the pre-conditioning stage is the availability of a natural resource to exploit as an economic lead sector, which might be timber, coal, oil or tropical crops. In this second stage it is still possible for the economy to stagnate or even slip backwards. This is why the third stage is so important: the third stage marks the economic take-off, during which self-sustaining growth is built into the economy so that, according to Rostow, it cannot regress.

The economic take-off requires the percentage of output reinvested each year to rise from less than 5 per cent to more than 10 per cent. This higher rate of investment would, with a reasonable level of efficiency of investment, raise economic output by more than 3 per cent per annum. A second characteristic of the economic take-off is the proliferation of groups in society capable of exploiting the new economic opportunities.

A third feature of the take-off stage is the emergence of an industrial lead sector such as textiles or steel with strong backward and forward linkages. In the case of steel, for example, backward linkage would take the form of the establishment of coal and iron ore mines to supply the steel mills. Forward linkage would comprise the factories set up to further process the steel output, such as shipyards and engineering works. In this way the industrial lead sector triggers the proliferation of new ventures throughout the economy. According to Rostow, the first take-off occurred in Britain in the

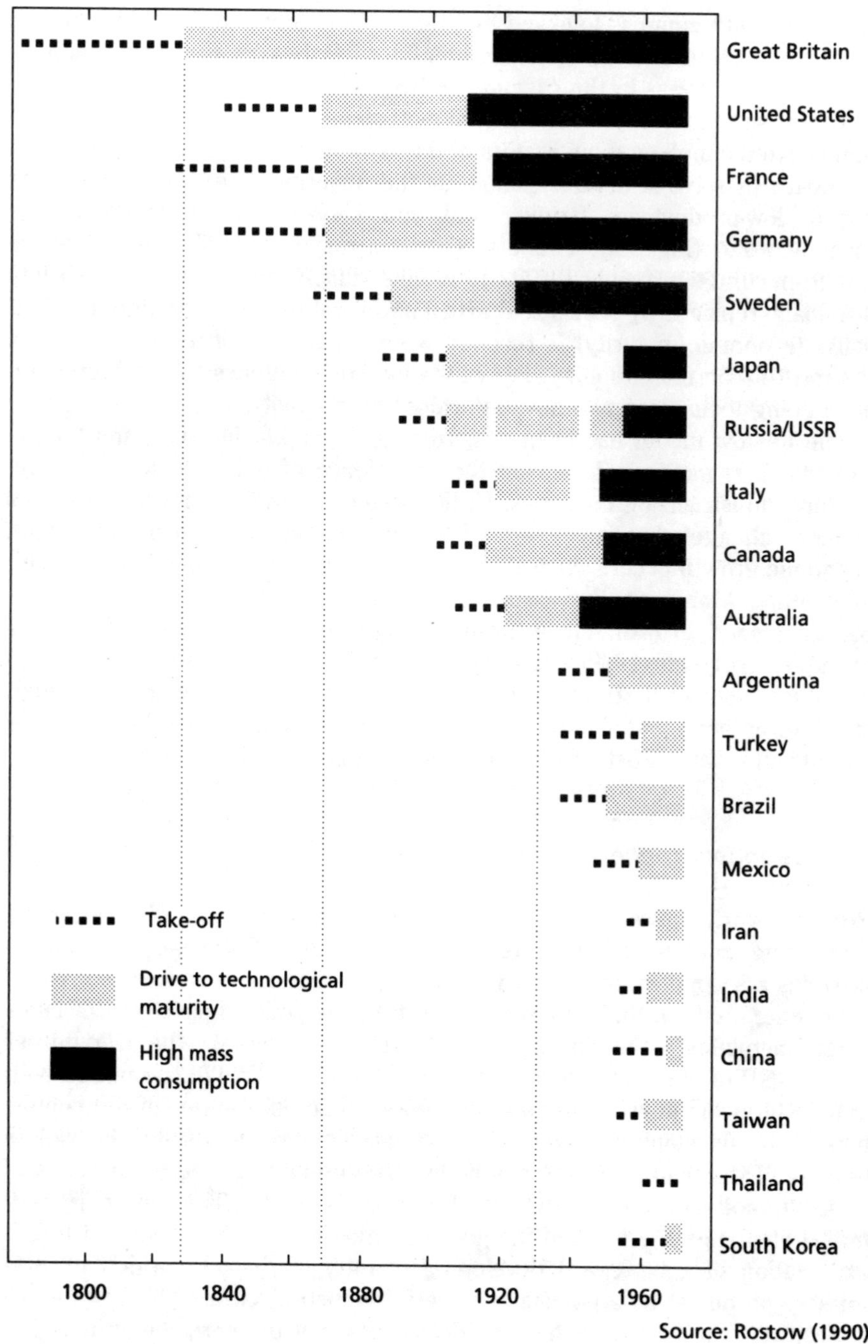

Fig. 1.1 *Rostow's five-stage model applied to twenty countries*

late eighteenth century, followed by France, Germany and the US in the late nineteenth century (Fig. 1.1). He identifies textiles as the lead sector in Britain, and railways in the case of the US.

The drive to maturity follows the take-off stage and involves the application of modern technology to all sectors of the economy – not just the lead sector. It implies massive structural change as the majority of the workforce shifts out of low-productivity farming and into services and high-productivity manufacturing (Fig. 1.2). This also entails a major redistribution of population from rural to urban areas; the look-alike village cells of traditional society are finally replaced by specialized urban and rural regions. The fifth and final stage (economic maturity) is reached when all sectors of the economy use the most modern technology. Per capita income continues to rise thereafter, as ongoing technological innovation raises productivity.

The Rostow model has been criticized for its right-wing bias, and for the fact that it is narrowly based on the experience of only fifteen nineteenth-century industrializing countries. Critics also claim that the model describes political changes but gives little insight into the mechanism by which economic growth occurs. In reality, the stages are often blurred and difficult to identify. Moreover, the assumption that all countries can – and will – pursue staged growth, albeit from a staggered start, is challenged. For example, Krugman (1979) suggests that the industrialized countries need to introduce a constant stream of new products, not just to maintain their high standard of living, but also to prevent themselves from slipping backwards—an outcome which Rostow's take-off stage suggests should not happen.

Core/periphery models

Rostow's work shows how easy it is to over-generalize when discussing the developing countries. The core/periphery model of the world economy provides a second example of a widely used model which carries the risk of over-generalization, this time from the left of the political spectrum. It takes global inequality as its starting-point. In 1990, for example, the average per capita GNP of the twenty-four OECD countries was $20 170, compared with only $840 in the developing countries. More than one-quarter of the population of the developing countries is estimated to live in a state of absolute poverty (Table 1.1), i.e. with insufficient income to meet the basic needs of housing, food and sanitation (World Bank 1990a, 1992a). The core/periphery model attributes the marked inequality in global income distribution to the exploitation of a backward developing country 'periphery' (which exports primary products) by a politically powerful industrial 'core'

Basically, the core/periphery model argues that the core region engages in unequal trade with the periphery so that it accumulates capital at the expense of the periphery. Myrdal (1957) portrayed the periphery as trapped in a series of vicious circles of poverty in which, for example, the initial

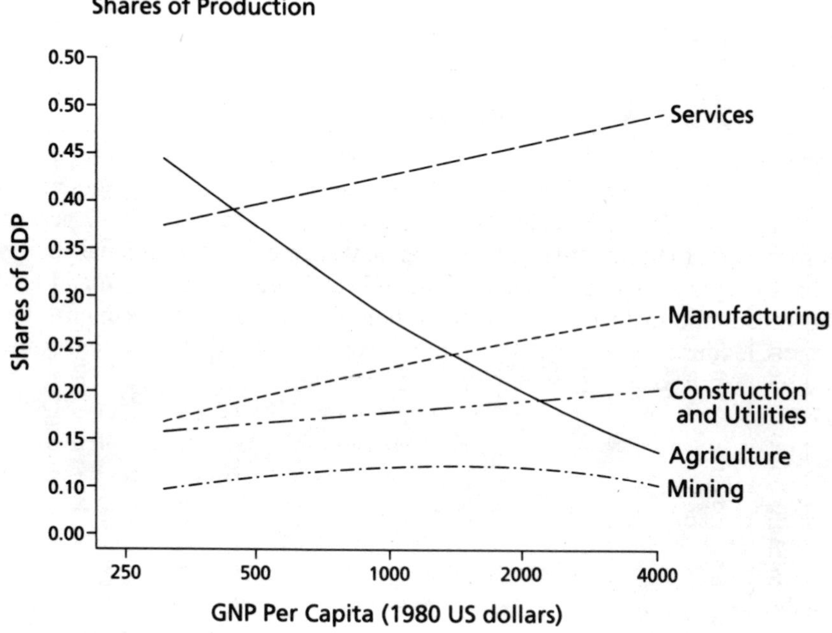

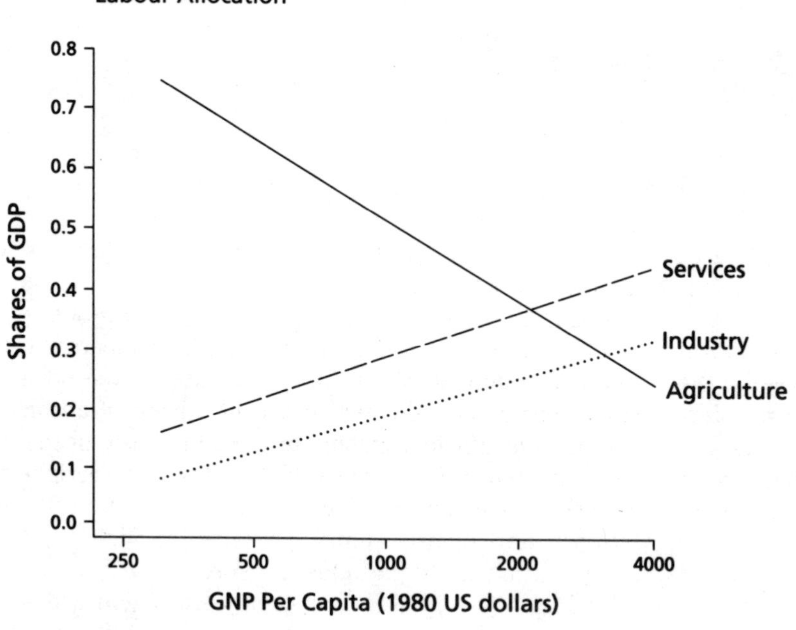

Source: Syrquin and Chenery (1989)

Fig. 1.2 *Per capita income and production structure*

Table 1.1 *Global distribution of poverty*

Region	Extremely Poor			Poor (including extremely poor)			Social indicators		
	Number (millions)	Headcount index (per cent)	Poverty gap	Number (millions)	Headcount index (per cent)	Poverty gap	Under-5 mortality (per thousand)	Life expectancy (years)	Net primary enrolment rate (per cent)
Sub-Saharan Africa	120	30	4	180	47	11	196	50	56
East Asia	120	9	0.4	280	20	1	96	67	96
China	80	8	1	210	20	3	58	69	93
South Asia	300	29	3	520	51	10	172	56	74
India	250	33	4	420	55	12	199	57	81
Eastern Europe	3	4	0.2	6	8	0.5	23	71	90
Middle East and North Africa	40	21	1	60	31	2	148	61	75
Latin America and the Caribbean	50	12	1	70	19	1	75	66	92
All developing countries	633	18	1	1116	33	3	121	62	83

Source: World Bank (1990a)

Note: The poverty line in 1985 PPP (purchasing power party) dollars is $275 per capita a year for the extremely poor and $370 per capita a year for the poor.

The headcount index is defined as the percentage of the population below the poverty line. The 95 per cent confidence intervals around the point estimates for the headcount indices are: sub-Saharan Africa, 19,76; East Asia, 21,22; South Asia, 50,53; Eastern Europe, 7,10; Middle East and North Africa, 13,51; Latin America and the Caribbean, 14,30; and all developing countries, 28,39.

The poverty gap is defined as the aggregate income shortfall of the poor as a percentage of aggregate consumption. Under-5 mortality rates are for 1980–5, except for China and South Asia, where the period is 1975–80.

absence of sufficient capital leads to low investment and therefore to low productivity. This in turn results in low income and therefore in low saving and inadequate capital accumulation.

Another formulation of the core/periphery model suggests that the growth-inducing trickle-down effects of trade with the advanced countries are modest compared with the negative backwash effects (the adverse impacts of such trade). The backwash effects debilitate the periphery and include the syphoning off of both capital and enterprising people. In addition, labour-intensive craft industries in the periphery decline because they are unable to compete with cheap imported mass-produced goods from industrial country factories. Structuralist reformers like Myrdal and Prebisch argue for state intervention to erect tariff barriers around developing countries, until they have built up a competitive manufacturing sector through the steady substitution of a growing range of domestic manufactured goods in place of imported manufactures. Membership of regional trading blocs like the Andean Pact would permit the capture of economies of scale because a single plant in one country could serve several national markets within the protected bloc.

Disappointment with the results of industrialization by import substitution led to a more radical interpretation of the core/periphery thesis in the 1960s. This took the form of the dependency school of thought associated with Gundar Frank, among others. In the view of this interpretation, North Atlantic capitalism had penetrated the newly independent Latin American economies from the mid-nineteenth century. Capital and technology from the industrial core were applied to extract a supply of raw materials for industrial processing and to simultaneously create new markets for industrial goods in the periphery.

Local elites in the peripheral countries were co-opted by the system, which concentrated the gains on the owners of capital, whether foreign or local, rather than upon the mass of the population. The economies of the dependent peripheral regions grew in response to the demand from the industrial core, so that their growth was not self-sustaining. If the industrial core should find cheaper alternative sources of materials, the dependent region would cease to grow and be left in a state of dependent underdevelopment. Gundar Frank cites poverty-stricken north-east Brazil as a classic example of a dependent economy. The dependency school rejects reform, in favour of revolutionary change in the social structure.

A later variant of dependency theory associated with Dos Santos, assigns a key role to the multinational corporations (MNCs). Dos Santos argues, with reference to Latin America, that the activities of MNCs represent the third stage of exploitation by the core, following on from the financial institutions of the late nineteenth century and the colonial governments before them. Yet another variant of dependency theory plays down the key role of external agents, concentrating instead on the attempts of local power groups to dominate the domestic political economy. This variant – internal colonialism (Walton 1975) – provides valuable insights into the political economy of post-colonial governments.

A common feature of core/periphery models is a deep pessimism concerning the present condition of and prospects for the great majority of the inhabitants of the developing countries. This contrasts with the optimism of orthodox observers like Rostow. In fact, the postwar experience of the developing countries reveals both successes and disappointments, so that a classification of developing countries is required in order to permit accurate generalization.

A country classification by resource endowment

The developing countries comprise more than 130 states, with very varied characteristics; in aggregate they account for more than three-quarters of the world's population, but generate barely one-quarter of its economic output. A classification of such countries into sub-groups will permit greater precision of analysis, allowing general statements to be qualified and making it possible to distinguish the features which all developing countries share from those which are peculiar to each class of developing country. Furthermore, comparisons of countries within each class permit the idiosyncrasies of individual countries to be identified and filtered out from shared group features.

Myint's three-country typology

There are clearly many criteria which could be used to classify the developing countries. This book opts for a system which, after Myint (1964), is based on the natural resource endowment, drawing primarily on country size. Myint recognizes three basic types of developing country, each of which faces a different set of constraints on its development. The three basic country categories are: large overpopulated, small overpopulated, and small underpopulated. The term 'underpopulated', as used by Myint, assumes that a larger population would be beneficial because it would enhance the prospects for capturing the scale economies in urban-based manufacturing and infrastructure. In addition, a higher population density would lower the cost of service provision in rural areas.

Myint's group of large overpopulated countries includes China, India, Pakistan and Bangladesh. These countries already faced intense population pressure on rural land even in the 1950s, when less than 20 per cent of their population was urbanized. This is because they lacked both the resource frontier enjoyed by many Latin American countries, and the ample land resources of many sub-Saharan African countries. For example, Hopper (1976) estimated that more than 83 per cent of the potential cultivable land had been used in Asia by the 1970s, compared with only 11 per cent in Latin

Table 1.2 *Contribution of expansion and intensification to cereal production increases 1961–90*

Country group	Production 1988–90 (million tons/yr)	Increase since 1961–3 (%) Total	Attributable to: Increased area	Increased yields	Current yield (tons/ hectare)
Developing Countries	1315	118	8	92	2.3
Sub-Saharan Africa	57	73	47	52	1.0
East Asia	499	189	6	94	3.7
South Asia	261	114	14	86	1.9
Latin America	105	111	30	71	2.1
Middle East and N. Africa	41	68	23	77	1.4
Former centrally planned Europe	336	76	–13	113	2.2
High-income countries	543	67	2	98	4.0
World	1858	100	8	92	2.6

Source: World Bank (1992a), p.135

America and 22 per cent in sub-Saharan Africa. Table 1.2 shows that the bulk of the increase in cereal output in Asia during the period 1961–90 came from increased yields, while Latin America and, especially, sub-Saharan Africa made sizeable acreage expansions.

Myint concludes that the large over-populated countries need to turn their large populations to advantage, building on the fact that they represent substantial home markets which facilitate efficient large-scale production. The appropriate development strategy for such countries is one based on the production of manufactured goods for domestic and export markets. The expansion of labour-intensive manufacturing absorbs labour from the land, so that those remaining in farming can acquire larger farms and thereby increase their incomes. Meanwhile, manufactured exports yield the foreign exchange with which to buy both the food which the country cannot produce efficiently for itself and the capital goods needed for industrial diversification.

The underpopulated group of countries accounts for many more individual nations than the large overpopulated group, but embraces fewer people. It includes much of Latin America, Africa and peninsular south-east Asia. Brazil and Mexico aside, the size of the domestic population in such countries is too small to support an efficient industrial economy. But these countries have ample land resources into which farming can expand. According to Myint, such countries should, therefore, develop via exports of primary products to the industrial countries. Industrialization will lag that of the large over-populated countries until the increased purchasing power of farmers offsets the small size of domestic markets and yields sufficient demand to support efficient manufacturing.

The third group of countries comprises the small overpopulated nations, including the island states of the Caribbean and Pacific, as well as Hong Kong and Singapore. The small resource-deficient countries of Taiwan and South Korea also fell within this group when Myint drew up his topology in the 1950s. Myint concludes that such countries face the direst prospect for development, because not only are they short of land, they also have domestic markets which are too small for efficient industrialization. Such disadvantages might, however, be offset by exploiting some natural resource such as minerals or sunny beaches. A second option is to encourage emigration. The third option is the most difficult because it requires the direct export of manufactured goods without the advantage of a relatively large captive home market. The fact that Myint considers the third option to be the greatest challenge makes the successful industrial development of the small resource-deficient Asian countries (the four 'dragons' of Hong Kong, Singapore, Taiwan and South Korea) the more remarkable.

A seven-country resource-based classification

Myint's 'underpopulated' category contains the great majority of developing countries. It can usefully be refined into five groups instead of one (Table 1.3). The revision involves:

- an income level split (into low and middle income);
- a cultural subdivision within the middle-income group (Latin American and South-East Asian);
- two groups, the mineral economy and the newly industrializing country (NIC), each of which is affected by a powerful modernizing export sector (mining and manufacturing, respectively).

The expanded classification has the advantage that it remains compatible with the widely used World Bank topology.

The separation of the low-income countries distinguishes those 'underpopulated' countries where the rural sector remains dominant from those (the mid-income countries) where it is not. Within the middle-income group, the Latin American countries have more severe constraints on their development as a result of foreign debt than their Asian counterparts. But already by the 1980s the three largest Latin American countries (Brazil, Mexico and Argentina) were being classified as NICs (Turner and McMullen 1982). The four Asian 'dragons' had already graduated into the NIC category.

The mineral economies account for almost one-quarter of the developing countries. They are defined as those developing countries which generate at least 40 per cent of their exports and 10 per cent of their gross domestic product (GDP) from minerals (Nankani 1979). Because of their mineral

Table 1.3 *A resource-based classification of developing countries*

Type	Income	Size	Population resources	Examples
Large lower income Asian	Low	Large	Overpopulated	India, China, Bangladesh, Pakistan
Small micro states	Varied	Small	Overpopulated	Maritius, Haiti, El Salvador
Small low-income sub-Saharan African	Low	Small	Underpopulated	Ghana, Kenya, Malawi, Niger
Small mid-income Latin American	Mid	Small	Underpopulated	Cuba, Costa Rica, Uruguay
Small mid-income SE Asian	Mid	Small	Underpopulated	Malaysia, Thailand, Philippines
Newly industrializing countries	Mid	Varied	Varied	Taiwan, S. Korea, Singapore
Mineral economies	Varied	Varied	Varied	Saudi Arabia, Zaire, Venezuela

Note: 1 Large countries have >30m population and >$30 billion GDP and >800 000 sq.km area.
2 Mid-income is defined as >$800 per capita income in 1985.
3 About one-tenth of the developing countries fall outside the above groupings (including Cambodia, Tunisia, Turkey and Vietnam).

bonus, they possess advantages for development that other primary product exporters lack. In particular, their mineral exports have the potential to provide a large source of foreign exchange and government revenues in addition to that furnished by agriculture. Moreover, mineral processing (or resource-based industrialization) offers an extra industrialization option. The mineral economies include both the oil-exporting countries like Venezuela, Nigeria and Indonesia, as well as ore-exporters such as Bolivia, Jamaica, Zambia and Papua New Guinea.

The economic problems associated with the two modernizing sectors (manufacturing and mining) are very different. Whereas the Asian NICs' export-oriented manufacturing is associated with greater economic flexibility and a high ability to cope with unexpected external shocks (Balassa 1982), a mineral export sector tends to amplify such shocks, often with detrimental socio-economic consequences (Auty 1993, Gelb 1988), as described in Chapter 9.

Both the NIC and the mineral economy groups cut across the global cultural regions. Cultural variations, however, remain a significant within-group differentiating factor: a recent survey of resource-based industrialization in eight oil-exporting countries found a deterioration in performance moving from East Asia and the Middle East to Latin America and sub-Saharan Africa (Auty 1990a). That contrast reflects differences in investment efficiency which are traceable to differences in economic policy.

Impact of the oil shocks on economic growth

The divergent responses of differing groups of developing countries to the 1973–4 and 1979 oil shocks reveal the policies most likely to lead to sustained and equitable economic growth. Each oil price increase transferred the equivalent of 2 per cent of GWP (gross world product) from the oil-importing countries to the oil-exporting developing countries. The oil-importing developing countries differed in their ability to adjust to this transfer.

Chenery (1981) estimated that with the prompt implementation of appropriate policies, the impact of the oil shocks would have been no more than a few months of slower economic growth. Such policies required the shift of capital and labour from activities oriented towards domestic consumption into production for export, in order to earn the extra foreign exchange with which to pay the higher costs of imported oil. In fact it cost most oil-importing countries (including many of the advanced countries) much more than this, because their policy response was confused and tardy.

Table 1.4 uses a World Bank classification of developing countries to trace the response to the oil shocks. It demonstrates that the large low-income Asian countries and the dynamic mid-income East and South-East Asian countries coped well and maintained relatively high rates of economic growth. The other three groups of developing countries were adversely affected by the oil shocks, including the oil-exporting mineral economies despite their higher export earnings. The difference in performance largely reflects differences in domestic policies.

Large low-income Asian countries

The satisfactory performance of the large low-income Asian economies, dominated by China and India, may be attributed to three factors. First, their large geographical area meant there was a greater likelihood that domestic energy resources would be available with which to substitute for high-cost imported oil. Second, as low-income countries with the vast majority of their workforce in farming, their economies were more sensitive to the performance of their agricultural sectors, which were benefiting from the rapid diffusion of green revolution techniques. Third, they pursued strongly autarkic (that is, self-sufficient) industrial strategies, in order to take advantage of their large domestic markets.

Table 1.4 *Differing responses to the oil shocks (% GDP growth/year)*

Countries	Industrial	Developing	Oil exporters		Oil importing developing countries					
					Mid-income				Low-income	
			High-income	Mid-income	E.Asia	Africa	Latin America	S.Eur	Asia	Africa
1960–73	5.0	6.0	10.7	7.0	8.2	3.5	5.6	6.4	4.6	3.5
1973–79	2.8	5.1	7.5	4.8	8.5	1.5	4.9	4.7	5.6	1.5
1980–85	2.2	2.9	–2.1	1.2	5.6	0.5	–0.2	1.7	6.3	1.1

Source: Gelb (1988)

The high degree of self-sufficiency of the larger low-income Asian countries reduced the share of trade in their GDP to a very small fraction (less than 5 per cent compared with over 40 per cent for smaller primary product exporters) so that they were strongly insulated against price shocks emanating from the international economy. Nevertheless, both India and China adopted reforms designed to open up their economies to international trade. In order to explain this apparent contradiction it is necessary to examine the second group of Asian economies, the vigorous mid-income countries of East and South-East Asia.

Dynamic mid-income Asian countries

A key feature of the dynamic Asian economies was their pursuit of an orthodox macroeconomic policy (World Bank 1993a). Such a policy has been associated with a competitive or open industrial policy (rather than a closed or autarkic one) and entails a twin commitment. The first commitment is to a competitive exchange rate (ensuring that the exchange rate does not rise above a level where exporters can no longer compete). The second commitment is to medium-term fiscal balance between government revenues and expenditures (to avoid the need for corrective action in the form of sharp public spending cuts or tax increases which deflate domestic demand and impede economic growth).

The net effect of such policies is to confer unusual flexibility on an economy: capital and labour can be rapidly shifted from areas of reduced demand to areas of high demand. More than most oil-importing countries, the dynamic Asian countries were able to respond to the oil shocks by quickly transferring productive resources from products for domestic consumption into the export goods required to pay for the higher cost of oil imports.

Moreover, the clear ability of such countries to use capital efficiently permitted them to borrow from abroad to bridge the initial external payments deficit caused by the higher cost of oil imports. The thrusting Asian countries did not, therefore, have to respond to the oil shocks by curbing non-oil imports, which would have slowed their economic growth. Finally, the competitiveness of their exports permitted these countries to comfortably repay their foreign debt during the 1980s.

Mid-income Latin American countries

In contrast to the dynamic Asian countries, the Latin American nations operated with less flexible economies. In part, this reflected the influence of structuralist reformers, such as Prebisch, who had urged them to build up their industrial sectors through import substitution in protected domestic markets. This entails temporarily raising external tariffs to make imports more expensive, thereby giving new domestic producers a breathing space within which to establish competitive production. The largest countries (Brazil and Mexico), like India and China (see Chapters 4 and 5), took advantage of their big domestic markets and diversified natural resource bases to pursue strongly autarkic industrial policies which stressed very high levels of self-sufficiency in manufacturing. Such policies create a slow-maturing manufacturing sector, in which firms may take decades to become internationally competitive.

The smaller Latin American countries also subsidized slow-maturing manufacturing sectors, despite their more limited domestic markets. They did so by using the rents from their minerals (i.e. revenues in excess of those needed by an efficient producer to remain in business) such as copper in the case of Chile and oil in the case of Venezuela, to provide the required foreign exchange and government revenues needed to support an uncompetitive manufacturing sector.

Most Latin American countries responded to the 1973–4 oil shock by borrowing abroad to accelerate the rate of import substitution. If the protected industries had matured rapidly, as they did in east Asia, such a policy might have worked. But the autarkic industrial policy favoured by Latin American countries continually extended subsidies and protection to manufacturing industry. Their uncompetitive manufacturing sectors could not adjust quickly to the deterioration in external conditions caused by an oil shock. Sachs (1985) shows that although the Latin American countries borrowed as much as the dynamic Asian countries, and had similar levels of public expenditure, their exports were much smaller in relation to their economies. Therefore, debt service payments during the 1980s absorbed a higher fraction of their export earnings than in the dynamic Asian countries (see Chapter 11).

Low-income sub-Saharan African countries

The countries of sub-Saharan Africa fared even worse than those of Latin America. Many had adopted inward-looking development policies after independence, in an effort to establish modern urban sectors. Like the Latin American countries, they favoured capital-intensive import substitution industry. They neglected farming, even though this was how the vast majority of Africans made their living and improvements there would have benefited the majority of the people.

Most countries of sub-Saharan Africa had lower credit ratings than those of Latin America, and so were forced to respond to the first oil shock by slashing their non-oil imports. This deprived their slow-maturing manufacturing sectors of the imported components they needed, and caused output to contract. Meanwhile, poor rural households were forced to substitute dwindling brushwood supplies for expensive imported kerosene fuel, accelerating soil erosion in some areas. The second oil shock merely intensified such problems, and its adverse economic effect was compounded by the debt crisis and declining commodity prices (see Chapter 2).

Oil-exporting countries

Each major regional group contains within it one or more oil-exporters, which Table 1.4 shows also experienced a deterioration in their economic performance after the oil shocks. Yet these countries were favourably affected by the oil price rises. For example, higher oil revenues meant that Nigeria received the equivalent of an extra 20 per cent of non-mining output each year from 1974 to 1984, as did Indonesia. Meanwhile, Saudi Arabian oil output was on such a scale in relation to its economy that the country received annually the equivalent of an additional 200 per cent of non-mining GDP in 1974–8, and 180 per cent in 1979–83. Yet economic growth in the oil-exporting countries decelerated sharply (Table 1.4). This is because, as Chapter 9 shows, their governments were seldom able to make efficient use of the extra resources they received. Political pressures caused an over-rapid domestic expenditure of the revenues. This expenditure triggered inflation and a strengthening of the exchange rate, which violated the twin requirements of orthodox macroeconomic policy.

The contrasting response to the oil shocks of the different types of developing countries underlines the importance of domestic policy. It also shows that those best-endowed with natural resources (the underpopulated countries of Latin America and sub-Saharan Africa, together with the oil-exporters) performed worse than those with a less bountiful resource base (such as the dynamic Asian countries). This suggests interesting links between key policy decisions and the natural resource endowment.

The resource curse thesis

The differential response of developing countries to the oil shocks suggests that the resource endowment may be inversely correlated with economic performance. In particular, the success of many resource-deficient East Asian countries contrasts with the disappointing outcome in much of resource-rich Latin America and sub-Saharan Africa. Not surprisingly, given this geographical pattern, explanations of differences in economic performance usually stress cultural factors (in the form of a work ethic or type of political regime) or environmental factors (the level of urbanization or the natural resource endowment).

The cultural thesis suggests that a Confucian factor accounts for the superior Asian performance (Morishima 1982). It argues that a respect for authority and a tolerance of hierarchy within Confucianism bring discipline and a greater readiness to sacrifice present consumption for future welfare in East Asian countries, compared with Latin America and sub-Saharan Africa. But the cultural thesis is flawed by the fact that there are many anomalies (Elkan 1978) such as the under-performance of the Philippines in South-East Asia, or the relative success of Colombia in Latin America. (Although it should be noted that the Philippines is a mainly Christian country rather than Confucian and that its history of Spanish colonialism suggests some likelihood of a degree of cultural affinity with Latin America.) Hughes (1988), for example, mischievously assigns the Philippines to Latin America and Colombia to Asia. Or again, Summers and Easterly (1992) point out that today's star performers are yesterday's basket cases, none more so than Korea and Singapore. Anomalies also pose similar problems for those such as Andrain (1988) and Bello and Rosenfeld (1990) who support the other important cultural thesis, which is based upon differing political regimes.

Sachs (1985) offers an 'environmental' explanation which attributes differences in economic success to variations in the degree of urbanization. He argues that high levels of urbanization foster the entrenchment of powerful industrial interest groups (firms and workers) which, once ensconced, favour the retention of import protection and oppose the adoption of orthodox economic policies. Protection of manufacturing raises domestic prices above those required by a globally competitive producer and creates rents (excess returns). The capture of such rents then gives the recipients a strong interest in perpetuating them. Sachs correctly notes that the successful East Asian countries had significantly lower levels of urbanization than did Latin American countries when key policy decisions were taken about the scale and nature of government intervention in the initial postwar decades (see Chapter 6).

Unfortunately, however, Sachs's comparison of countries in South-East Asia and Latin America is too narrow: other countries with low levels of urbanization in Asia, as well as in most of sub-Saharan Africa, failed to prevent the entrenchment of such interest groups. Yet although Sachs has selected the wrong causal factor in picking the level of urbanization, the entrenchment of powerful urban rent-seeking groups does play a critical role in blocking the reform of errant policies, once they have been adopted. But a second environmental factor, the natural resource endowment, may account for differences in policy choice.

The 'resource curse' thesis suggests that resource-rich countries may squander their resource advantage because an over-optimistic estimate of their prospects leads to the pursuit of lax economic policies. A corollary is that resource-poor countries, mindful of their marginal position, may compensate for their disadvantage by adopting firmer and more far-sighted policies.

Taking country size as one aspect of a beneficial resource endowment, a recent examination of the effect of country size on economic performance (Perkins and Syrquin 1989) shows that the industrial structure of larger countries tends to diversify early and that large countries tend to be more self-sufficient or autarkic. Yet, as noted earlier, small resource-rich countries like the mineral exporters may also develop a large protected manufacturing sector which must be subsidized from the rents on commodity exports such as minerals or cash crops.

More generally, a bountiful natural resource endowment may counter-intuitively affect policy negatively in three ways. First, it may lead initially to tolerance of lax macroeconomic policies: for example, through an over-valued exchange rate which cheapens imports but also saps the competitiveness of the unprotected activities such as light industry and agriculture; or through persistent budget deficits based on over-optimistic estimates of future resource revenues. Second, over-optimistic projections of resource earnings also ease pressure to develop competitive manufacturing. Third, the persistence of macroeconomic imbalances (basically the trade and budget deficits) and failure to reform manufacturing (in the face of pressure from those groups which benefit from industrial protection) makes change increasingly difficult.

If the attainment of industrial competitiveness is delayed, as is the case with an autarkic industrial policy, then the burden of generating foreign exchange remains with the shrinking primary sector (Fig. 1.2). A resource-rich country can sustain the slower industrial maturation for longer than a resource-deficient country, because its primary sector generates much greater rents. But as per capita income rises and the relative size of the primary sector declines, the manufacturing sector is eventually required to compensate (Fig. 1.2). A weak manufacturing sector is ill-placed to do so. Moreover, the longer it has been protected, the more resistant to change become the interest groups (such as the workforce and owners of protected industry) which benefit from protection. They block reform, so that in this way a rich natural resource endowment can trigger a set of policy choices which, within little more than a decade (Gelb *et al.* 1986), can transform the resource bonus into a curse.

Development policy and economic performance

Pessimistic observers of postwar economic development have tended to attribute slow economic growth and inequitable income distribution to the process of capitalist development. But the inter-country comparisons of economic performance do not support such interpretations: rather they indicate that disappointing development results from the cumulative impact of incorrect economic policies. In particular, the experience of the East Asian dragons (notably Korea and Taiwan, whose larger area and domestic markets makes them of more general interest than the city states of Singapore and Hong Kong) may be summarized in terms of a four-stage model (Kuznets 1988). Basically, Taiwan and Korea both embraced autarkic industrial policies

Table 1.5 *Investment efficiency and GDP growth by policy phase, Taiwan and Korea*

Policy phase	Autarkic industrial policy	Competitive industrial policy Export-led	HCI drive	Liberalized
Dates: Taiwan	1950–8	1959–71	1972–81	1982–91
Korea	1952–62	1963–73	1974–82	1983–91
Investment (%GDP)				
Taiwan	12.2	17.2	27.4	21.4
Korea	n.a.	19.9	29.1	30.9
Incremental capital output ratio				
Taiwan	1.5	1.8	3.0	2.6
Korea	n.a.	1.9	4.1	3.0
GDP growth (%/Yr)				
Taiwan	7.9	9.5	9.1	8.2
Korea	n.a.	10.4	7.1	10.2

Source: Auty (1994, p.17)

in the 1950s but they both abandoned this strategy in favour of a competitive industrial policy, which they pursued within the context of a pragmatically orthodox macroeconomic policy.

There is clear evidence that an autarkic policy damages long-term economic performance, whereas a competitive industrial policy may well be highly beneficial (Amsden 1989, Wade 1990), although that view is contested (Ranis and Mahmood 1992), as is shown in Chapter 11. A competitive industrial policy staves off foreign competition only temporarily. It simultaneously maintains the competitiveness of established industrial sectors while causing the rapid emergence of new manufacturing sectors (Ohno and Imaoka 1987). Under a competitive industrial policy, governments provide a package of market information, assistance with technology acquisition, subsidized credit, tax breaks and trade incentives for new firms to set up 'infant' industries. But, unlike an autarkic policy, the government demands that favoured firms rapidly achieve international competitiveness.

Following their initial autarkic phase of industrialization (often referred to as primary import substitution), Taiwan and Korea industrialized in three distinct stages (Table 1.5). Stage two stresses labour-intensive exports, before the emphasis shifts to the promotion of heavy industry in stage three (the process of secondary import substitution). The expansion of labour-intensive manufactures (much for export) in stage two rapidly absorbs surplus labour so that the 'turning point' is reached, which in the case of

Taiwan and Korea occurred around 15 years after the abandonment of autarkic policies.

The turning point is reached when the rapid expansion of labour-intensive light industry has absorbed surplus domestic workers, so that labour shortages occur. This triggers pressure for wage increases which must be accommodated by raising worker productivity (Kuznets 1988). In both Taiwan and Korea the rapid absorption of surplus labour out of the rural sector was assisted by weak unionization, which kept wages low and labour flexible. This is one of the less appealing features of the model (Bello and Rosenfeld 1990), but it does ensure the broad participation of the workforce in the development process, one of two necessary conditions, according to the World Bank (1990a), for poverty alleviation. (The second condition is the careful targeting of government assistance on those most in need.)

In the case of Korea and Taiwan, the rapid expansion of workforce participation in the modern economy built on two favourable preconditions, namely land redistribution and the provision of primary education, to maintain a relatively equitable income distribution. This expansion contrasts sharply with the creation of the large unemployed or underemployed 'marginalized' class characteristic of Latin America and sub-Saharan Africa. In these countries, most governments attempted to leap-frog the second (labour-intensive export) stage of the East Asian development model.

The ratio of the income of the richest quintile to that of the poorest quintile was around 4 in Taiwan and 7 in Korea, compared with 33 in Brazil and 23 in Kenya (Sachs 1989). In addition to relatively equitable income distributions, Taiwan and Korea both achieved rapid economic growth. Over the period 1961–90, GDP growth in Taiwan averaged 9.1 per cent compared with 8.7 per cent in Korea (Table 1.5). But progress in combating environmental pollution and in achieving political freedom has been rather slower to emerge in these countries (Bello and Rosenfeld 1990).

The turning point heralds the third stage of the East Asian development model, which sees the gradual abandonment of lagging, low-productivity manufacturing like simple assembly operations to the next wave of NICs and the expansion of high-productivity sectors of emerging competitive advantage like steel, electronics and increasingly sophisticated engineering. Whereas the initial phase of a competitive industrial policy is strongly export-driven, domestic demand becomes a steadily more important source of economic growth in stage three, as rising wages boost domestic purchasing power (Chenery *et al.* 1986). Stage four sees the expansion of research-intensive activity and a decline in state intervention, as the economy approaches industrial country standards of income and complexity.

The entire sequence traced out by the East Asian development model can propel a country from low-income to industrial status in little more than one generation. But it calls for the adoption of and adherence to sound economic policies, a condition which a constrained natural resource endowment appears more likely to elicit than a bountiful one.

Conclusion and structure of the book

The difficult conditions created by the oil shocks of 1973 and 1979 provided a test of economic resilience which confirmed the superiority of the orthodox approach to economic management for both economic growth and income distribution. Williamson (1993) has termed this outcome the 'Washington consensus'. But within the broad consensus, controversy remains over important issues such as the role of the state (in industrial policy and other matters), the policies required to achieve sustainable development and the degree of political freedom compatible with rapid economic transformation.

There is also emerging evidence that the choice of macroeconomic and industrial policy is linked to differences in the resource endowment of countries. The poorer economic performers have tended to be the more favourably endowed countries. Although such a conclusion lacks the strength of a law (as is clearly illustrated with reference to Indonesia and Malaysia in Chapter 10), it does represent a strong tendency. Well-endowed countries like resource-rich and market-rich Brazil and Mexico, as well as most oil-exporting and ore-exporting countries, appear more likely than resource-deficient countries to make policy errors. Basically, it appears that a rich resource endowment becomes a curse by giving well-endowed countries more latitude to pursue less prudent development policies.

The well-endowed countries have attempted to use their resource advantage to accelerate the process of industrialization, leap-frogging the second stage of labour-intensive exports in the East Asian development model and advancing prematurely into heavy industry. The result has been a slow-maturing manufacturing sector which is dependent on the primary sector for foreign exchange and subsidies. As the latter sector declines in relative size with rising per capita income, the burden imposed on other economic sectors by the manufacturing sector becomes too great. Economic growth becomes more erratic and slower, while income inequality is amplified as inadequate employment opportunities create a marginalized class which benefits little from the process of economic growth. Yet policy reform is difficult to achieve in the face of vested interest groups which benefit from a protected economy.

How these broad relationships between resource endowment, policy choice and economic performance work out in specific types of country is explored through the remainder of this book. Each section of the book tackles a general set of problems of particular relevance to a subset of developing countries.

Part 2 focuses on population pressure and land resources with particular reference to Africa, the global region which has the highest rate of population growth and a majority of its population still engaged in agriculture. The urban bias of past policies is examined and the consequences of the resulting rural neglect for living standards and environmental damage are reviewed for the arid and semi-arid regions of the continent, where population and land pressures are often at their most acute. An alternative strategy for sustainable African development is discussed in Chapter 3.

The role of agriculture in low-income countries remains a key theme in Part 3, which in Chapter 4 analyses the prospects for reducing income inequality through policy reform, with particular reference to India and China. Attention then turns in Chapter 5 to the problems of the autarkic industrialization policies espoused by the two Asian giants. The difficulty of achieving reform, even when the failings of the original policy are widely appreciated, is explained and the emerging regional inequalities in China are discussed.

Part 4 turns to the mid-income countries of Latin America, where a majority of the population is already urbanized and mistaken policies have left a legacy of high income inequality. The themes of hyper-urbanization (the over-rapid expansion of very large cities) and marginalization (the creation of a large, low-income workforce outside the formal economy) are examined in Chapter 6. The planners' attempts to resolve the problems of urban congestion by urban restructuring and regional policies are evaluated in Chapter 7.

The counter-intuitive consequences of a rich natural resource endowment provide the theme of Part 5, which also shifts the focus to global core/periphery relationships. It examines the efforts of a fourth set of developing countries, the mineral exporters, to restructure the global economy and establish a new international economic order. The attempt by the OPEC cartel to set oil prices is one such example, and Chapter 9 traces the impact of the 1973 and 1979 oil shocks on both the potential gainers from higher oil prices (the oil exporters) as well as the losers (the oil-importing ore exporters). Chapter 10 demonstrates, however, that the resource curse thesis is not a deterministic law, by examining the rapid development of resource-rich Indonesia and Malaysia.

Part 6 of the book turns to the successful Asian NICs which, with relatively poor natural resource endowments, embraced the international economy and turned it to their advantage. Their success makes the role of governments in promoting rapid, equitable and sustainable growth an appropriate theme for Chapter 11. The concluding chapter of the book draws policy lessons for future global development.

PART 2

Urban bias and environmental deterioration: low-income Africa

2

Rural neglect and environmental degradation in Africa

False start in Africa

In a perceptive book, translated into English in the mid-1960s, the French economist René Dumont (1966) argued that the newly independent countries of Africa had made a false start on the development path. His insight is the more remarkable, because he wrote at a time when many African countries had recently gained their independence and most were basking in a wave of optimism concerning the prospects for post-colonial society. Yet over the next three decades economic advance proved disappointing. After a promising start in the 1960s, per capita economic growth decelerated, while per capita food production also declined. This meant that poverty increased, especially in the rural areas where a majority of the population continued to live. In the late-1980s farming still provided three-fifths of employment, two-fifths of exports and one-third of GDP in sub-Saharan Africa (Cheru 1992).

Opinion is divided as to the reasons for this disappointing economic performance. The divide broadly parallels that outlined in the previous chapter between orthodox and structuralist observers. More precisely, structuralists attribute the disappointing progress predominantly to external causes (such as deteriorating export commodity prices, negative oil price and interest rate shocks, and onerous foreign debt) while the orthodox school considers that internal factors (domestic policies) are fundamentally to blame. But, whatever the cause, the consequences for the rural environment and the people who depend most closely upon it have been very damaging in some regions of Africa.

This chapter explores the reasons for this disappointing start and outlines the consequences with reference to some of the arid and semi-arid regions of the continent (Fig. 2.1), where the problems tend to be most acute. The semi-arid region occupies 42 per cent of the land area and contains two-thirds of the people of sub-Saharan Africa (Toulmin *et al.* 1992). It stretches for 4500 km, through the Sahel of West and Central Africa (Fig. 2.2) before sweeping round the Horn of Africa and extending south into Kenya, Tanzania and eventually Southern Africa.

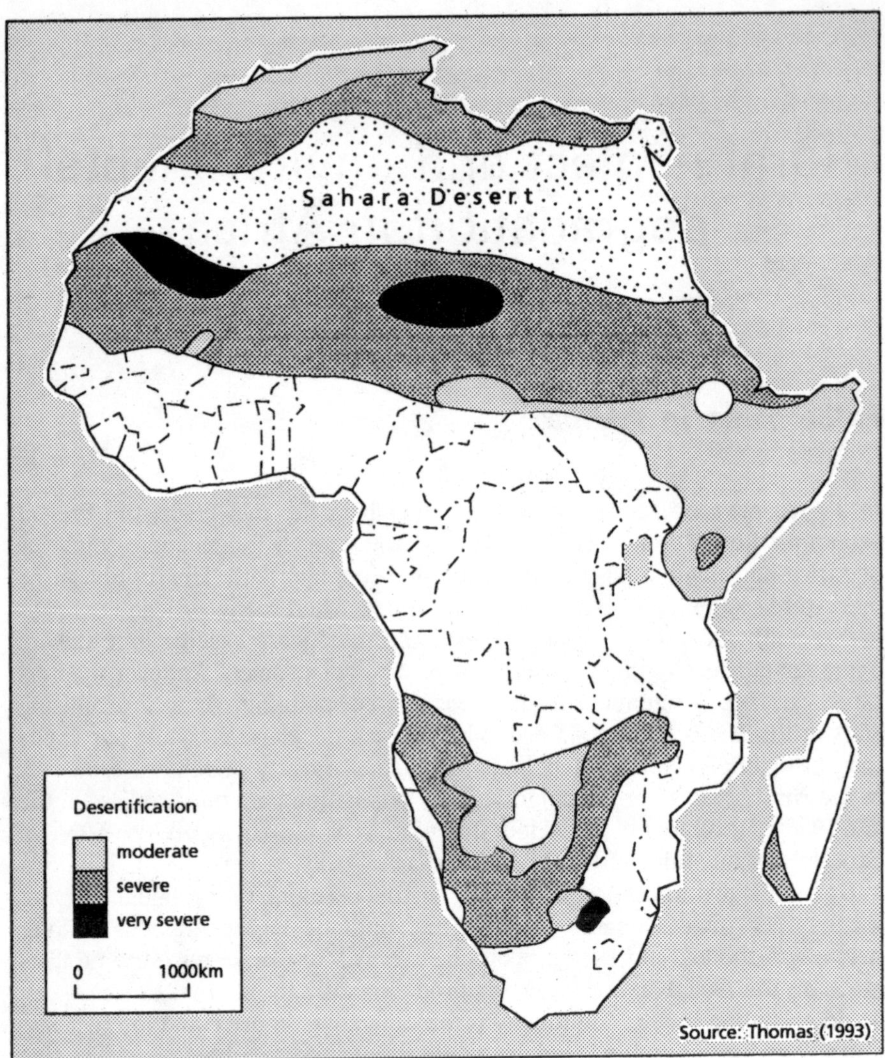

Fig. 2.1 *Arid and semi-arid zones of Africa*

The chapter begins by evaluating the relative importance of external and internal factors in accounting for the region's disappointing rate of economic growth. It is argued that although external conditions have not been favourable for most African countries, domestic policy failures turned a difficult situation into a crisis. Such a conclusion is consistent with the orthodox approach, but that approach has rather less to say about why such inappropriate policies were adopted. Instead, the structuralists have the more useful insights, most notably in the form of internal colonialism, which is examined

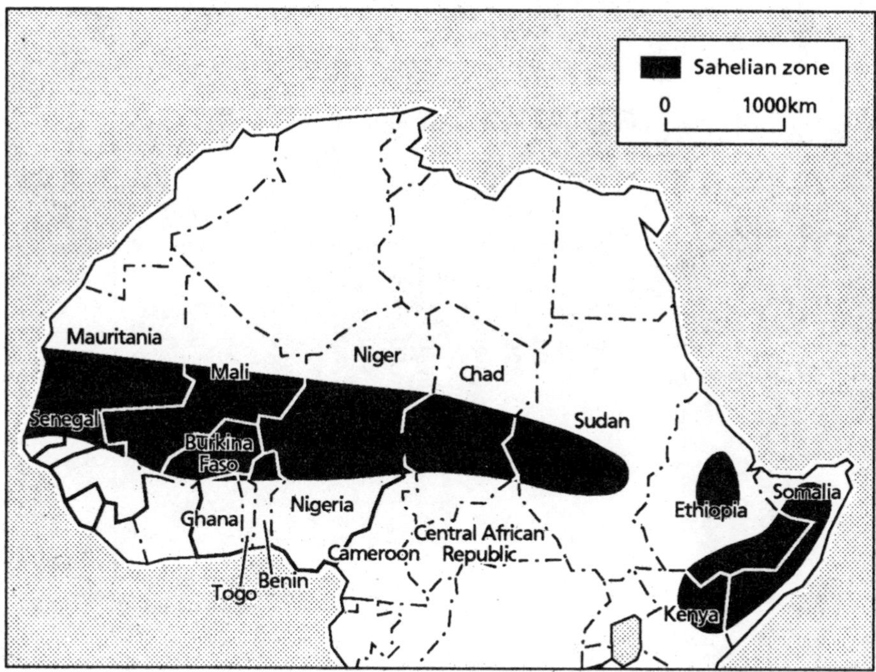

Fig. 2.2 *The Sahelian zone of Africa*

in the second section of this chapter. It may be recalled from the previous chapter that internal colonialism is a political system in which a largely urban-based national elite uses its political dominance to squeeze revenues from the rural majority. Such an outcome echoes the urban bias thesis of Lipton (1977) and can be seen in other types of country, not least the low-income Asian countries examined in Chapter 4.

The third section of the chapter traces the consequences of poor economic performance for the environment of the semi-arid regions. It demonstrates the critical role of technology in mediating between rising population and land resources. But section four cautions against resorting to a large 'technological fix'. It does so with reference to water resource management in the lower Nile Valley. This sets the stage for an analysis in Chapter 3 of the role of technology in rural economic development and, in particular, of the relationship between technology and the relative scarcity of land.

Post-independence economic growth

This section focuses on the sub-Saharan African countries, excluding southern Africa and the Maghreb (Table 2.1), where differences in rates of

Table 2.1 *Economic and social indicators for selected African countries*

Country	Area (million sq. km)	Population (million)	Cropland/ hd (ha)	Population growth (%/yr)	Population in cities 1991 (%)	GDP growth 1980–91 (%/yr)	Per capita Income ($1991)	Literacy (%)	Life expectancy (yrs)
North East									
Egypt	1.001	53.6	0.05	3.8	47	4.8	610	48	61
Ethiopia	1.222	52.8	0.28	3.1	13	1.6	120	n.a.	48
Kenya	0.580	25.0	0.10	2.5	24	4.2	340	69	59
Tanzania	0.945	25.2	0.19	3.0	34	2.9	100	n.a.	51
Uganda	0.236	16.9	0.36	2.5	11	n.a.	170	48	46
Sudan	2.506	25.8	0.50	2.7	22	n.a.	n.a.	27	51
West									
Chad	1.284	5.8	0.56	2.4	30	5.5	210	30	47
Ivory coast	0.322	12.4	0.31	3.8	41	(0.3)	690	54	52
Ghana	0.239	15.3	0.18	3.2	33	3.2	400	60	55
Mali	1.240	8.7	0.23	2.6	20	2.5	280	32	48
Senegal	0.197	7.6	0.71	3.0	39	3.1	720	38	48
Sierra Leone	0.072	4.2	0.43	2.4	33	1.1	210	21	42
South									
Malawi	0.118	8.8	0.28	3.3	12	3.1	230	n.a.	45
Madagascar	0.587	12.0	0.26	3.0	25	1.1	210	80	51
South Africa	1.221	38.9	0.37	2.5	60	1.3	2560	n.a.	63
Zimbabwe	0.391	10.1	0.29	3.4	28	3.1	650	67	60

Sources: World Bank (1993b), WRI (1992)
Note: See also Table 9.1 for data on some African mineral economies

Table 2.2 *Economic growth, by major developing regions, 1965–89*

	Growth of real per capita GDP(%)			Gross domestic investment/GDP(%)		
	1965–73	1973–80	1980–9	1965–73	1973–80	1980–9
Sub-Saharan Africa	3.2	0.1	–2.2	16.2	20.8	16.1
East Asia	5.1	4.7	6.7	24.2	29.7	30.0
South Asia	1.2	1.7	3.2	17.1	19.9	22.3
Eastern Europe	4.8	5.3	6.8	28.3	33.8	29.4
Middle East and N Africa	5.5	2.1	0.8	23.4	29.2	25.9
Latin America and Caribbean	3.7	2.6	–0.6	20.7	23.9	20.1

Source: World Bank (1990a)

decolonization and resource endowment have resulted in more varied outcomes (Fig. 2.3). Explanations for the disappointing economic performance of sub-Saharan Africa fall into two classes, depending upon whether the causes are believed to be predominantly external or internal. The external arguments begin with the colonial legacy of inadequate education and an infrastructure geared to the extraction of minerals and cash crops (Acharya 1981). Yet, prior to the first (1973) oil shock, the economic performance of sub-Saharan Africa was quite respectable: the rate of per capita economic growth averaged 3.2 per cent annually during the period 1965–73, outstripping the growth rate of South Asia by a considerable margin (Table 2.2).

A second external factor invoked to explain the disappointing performance of sub-Saharan Africa is the long-term deterioration in the export prices of most African commodities. This trend was exacerbated first by growing price instability in the 1970s and then by crippling levels of external debt. Per capita GDP in sub-Saharan Africa stagnated between 1973 and 1980, and then contracted by more than 2 per cent annually during the 1980s. But the deterioration in exports did not lead to a harmful decline in investment rates. Certainly, the rate of investment, having climbed to a respectable one-fifth of GDP between 1973 and 1980, fell back during 1980–9. But the rate of investment was no lower in the 1980s than it had been between 1965 and 1973, when economic growth was satisfactory. It is therefore not a lack of investment which explains sub-Saharan Africa's slowing economic growth, but rather the declining efficiency of that investment.

The deterioration in investment efficiency cannot be explained by the disappointing prices for African exports. Although the prices of African exports did fall in real terms relative to the price of (mostly imported) manufactured

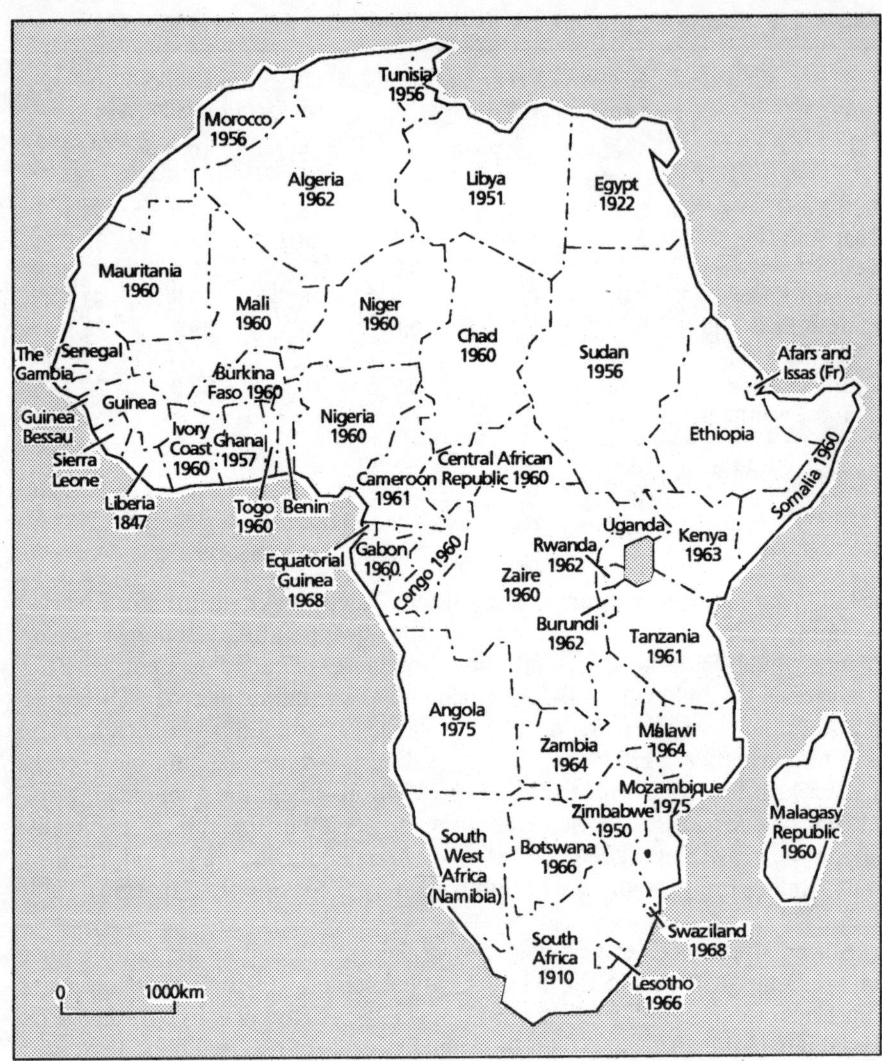

Fig. 2.3 *Attainment of independence by African countries*

goods, not all sub-Saharan African countries were adversely affected. Killick (1992) notes that although the oil-importing countries of sub-Saharan Africa were adversely affected by deteriorating export prices, two-fifths of the region's exports comprise oil (World Bank 1992b). More specifically, important oil-exporters like Nigeria and Cameroon experienced improvements in export revenues, but their economic performance was also disappointing (Auty 1990a).

Table 2.3 *Distribution of poverty by region, 1985*

Region and country	Rural population as % total	Rural poor as % total	Infant mortality (Per 1000 live births)		Access to safe water (% population)	
			Rural	Urban	Rural	Urban
Sub-Saharan Africa						
Ivory coast	57	86	121	70	10	30
Ghana	65	80	87	67	39	93
Kenya	80	96	59	57	21	61
Asia						
India	77	79	105	57	50	76
Indonesia	73	91	74	57	36	43
Malaysia	62	80	n.a.	n.a.	76	96
Philippines	60	67	55	42	54	49
Thailand	70	80	43	28	66	56
Latin America						
Guatemala	59	66	85	65	26	89
Mexico	31	37	79	29	51	79
Panama	50	59	28	22	63	100
Peru	44	52	101	54	17	73
Venezuela	15	20	n.a.	n.a.	80	80

Source: World Bank (1990a)

The World Bank (1989a) calculates that even setting aside the oil-exporters, the more unfavourable export prices which sub-Saharan African countries experienced compared with other low-income countries, should not have depressed the economic growth rate by more than a half per cent annually. An important reason why other low-income countries encountered better trading opportunities than sub-Saharan Africa is because they diversified away from export products with declining prices (Duncan 1993). Thus, whereas all low-income countries reduced the share of primary products in their exports from 76 per cent to 48 per cent between 1965 and 1989, sub-Saharan Africa moved only marginally, from 92 per cent to 89 per cent. Expressed another way, most sub-Saharan African countries were very slow in adjusting to declining export prospects.

As for another important 'external' factor, the burden of foreign debt, the countries of sub-Saharan Africa borrowed on far more favourable terms than those of Latin America. For example, the period over which they were required to repay their loans was twice as long as on Latin American loans, while interest rates were half those on Latin American loans. Moreover, sub-Saharan Africa also receives a relatively high share of global development aid, much of it in the form of outright grants. Under these circumstances, the inability of sub-Saharan African countries to finance their debt lies less with the terms of the loans and more with the region's sluggish response to falling export prices.

Poor policies have meant that the incidence of poverty in most of the region has been getting worse, in both absolute and relative terms. By 1985, almost half of sub-Saharan Africa's estimated population of 382 million was classed by the World Bank as poor. This is a similar level of poverty to South Asia (51 per cent), but far higher than East Asia (including China) and Latin America where around one-fifth of the population is classed as poor (World Bank 1990a). The policies which most sub-Saharan African countries have pursued since independence have discriminated against farmers; as a result, poverty is disproportionately concentrated in rural areas (Table 2.3).

Internal colonialism and urban bias

Dumont (1966) perceptively argued that a new elite had replaced the colonial elite in sub-Saharan Africa, and that the new elite placed too much emphasis on the rapid creation of a modern urban infrastructure and industry. A corollary is that the rural areas, where 90 per cent of Africans then lived, were being neglected. The World Bank (1989a) estimates that during the first two post-independence decades, some two-thirds of all investment in sub-Saharan Africa went into cities. That high rate of urban investment, and neglect of the rural economy, lies at the root of the decelerating economic growth and rising poverty in sub-Saharan Africa.

Dumont's insight is consistent with the concept of internal colonialism, an idea developed by the structuralist school of development theory. It is succinctly defined by Walton (1975) as the domination by a privileged minority of a dependent majority. Internal colonialism, unlike dependency theory, does not assign the prime responsibility for internal patterns of exploitation to external agencies, although it does view the elite as using external trade as a means of accumulating capital.

The urban elite uses its political power to extract surpluses at the expense of the rural poor in four ways. First, the provision of infrastructure is designed overwhelmingly to facilitate the exports from which surpluses are extracted. Second, credit and other inputs are provided for large-scale export-oriented farms at the expense of small-scale producers of crops for the domestic market. Third, domestic food prices are set low in order to reduce urban wage costs. Fourth, tariffs are used to protect capital-intensive (low-employment) domestic industry. Within this inefficient manufacturing sector, high-cost goods are produced by a well-paid labour aristocracy for consumption by the elite and the labour aristocracy. The system, in effect, taxes domestic food producers in order to create industrial and government employment for a favoured minority in the cities.

Lipton (1977) argues that in low-income countries the return on investment in agriculture is potentially two to three times that in other sectors. But urban interests inevitably win the political struggle with the rural poor. This is so, despite the fact that the countryside contains a majority of the population, most of the poverty and most of the low-cost opportunities for economic advance. An alliance between urban groups and larger farmers ensures that the latter gain access to cheap credit and subsidized farm inputs. The larger farmers will sell what they produce to the cities, whereas small farmers are initially likely to consume much of their extra production themselves. By favouring larger farmers, a supply of cheap food is made available to the cities where most economic activity is shielded from foreign competition, remaining relatively unproductive and inefficient. Such a situation cannot continue indefinitely: policy reform is crucial if growing poverty and severe environmental damage are not to result from the lethal combination of population growth and economic stagnation.

The internal colonialism and urban bias theses are in line with the new political economy which sees governments not as maximizers of long-term national welfare, but as arbiters between competing pressure groups. They are also consistent with the notion of rent-seeking behaviour as expounded by the neo-liberal wing of the orthodox school of thought. In the neo-liberal view, the result of state intervention in the economy is all too often the undesirable creation of rents (returns in excess of those required by a competitive world producer) for distribution among privileged groups. Such redistributive policies can seriously undermine the overall efficiency of investment in an economy, and within little more than a decade.

For example, Gelb *et al.* (1986) show how attempts by a newly independent African government to raise urban wages towards expatriate levels proved counter-productive. Food prices were kept low and rural incomes and incentives were further depressed by the tight central control (and inept administration) of agricultural incentives. Not surprisingly, with such disappointing opportunities in peasant farming, rural–urban migration intensified, and government fear of urban unrest led to the creation of unnecessary jobs (overmanning) in both government offices and state-owned commercial firms alike, further impairing efficiency and competitiveness.

A vicious circle was created which proved capable of depressing the efficiency of national investment within a decade to levels where economic growth was below the rate of population increase. Worse, in impoverished rural societies, where the priority becomes the need to ensure immediate survival, environmental conservation receives low priority. The economic deterioration in many African countries after the 1960s has been accompanied by environmental degradation in some regions.

Economic deterioration and environmental pressures in the Sahel

The impact of drought on the pastoral zone

The Sahel proper comprises a 4500-km strip running from Senegal in the west through Mali and Burkina Faso to Niger and Chad (Fig. 2.2). The sub-region derives its name from the Arab word for coastline, a reference to its position at the southern edge of the Sahara desert. Within the Sahel rainfall averages 250–600 millimetres/year, with a short rainy season around February and a longer rainy season lasting from June to September. The economy of the Sahel traditionally formed part of a tri-regional symbiosis with the savanna dry-farming zone and the rainforest zones to the south. But as the population of West Africa increased, and tastes changed in the major urban markets, that symbiosis broke down (DuBois 1974) and pressure on the rural environment intensified.

The Sahel is the home of nomadic headers of cattle and goats, such as the Fulani and Taureg. Their communal grazing rights bar individual or cooperative ownership, and this reduces incentives to invest to conserve grazing land and water resources. This is because, as Hardin (1968) argued in 'The Tragedy of the Commons', any reduction in herd size by an individual, designed to ease the pressure on grazing, is negated because other pastoralists expand their herds to take advantage of the extra grazing. In any case, such an individual reduction in herd would be unlikely to occur because the herd size has social significance – high numbers symbolize wealth. A large herd is also seen as insurance against drought, because individual animals can be sold off to secure food to tide the owner over a period of hardship without threatening the ability to rebuild the herd size under better times.

The Sahelian herders rotated their animals between pastures and wells. They traditionally migrated south to sell their livestock in the savanna region and coastal cities up to 1000 kilometres away. For example, livestock exports comprised between two-thirds and four-fifths of export earnings in Chad and Niger in the late 1960s. As the herders passed through the savanna zone *en route* for the urban markets, they provided seasonal labour to the dry-farmers at harvest time. Rainfall is both higher and more regular in the savanna zone, and farmers there plant millet and sorghum in the April short rains, which grow rapidly with the high insolation and longer rainy season. In return for their assistance with the harvest, the pastoralists were allowed to graze their herds on the stubble in harvested fields, which the animals manured as they grazed. After the harvest, surplus animals and grain moved south to the urban markets, at a rate estimated at 700 000 head annually in the late 1960s.

This symbiosis between the Sahelian pastoralists, the savanna farmers and the southern urban markets began to break down as numbers and incomes rose in the southern cities and diets began to change. The new urban dwellers demanded a more varied diet, with higher protein content. They showed an increasing preference for convenience foods such as rice and wheat over grains such as millet from the savanna dry-farm zone and the traditional root crops of the forest zone. This changing urban demand had two important effects on the regions to the north. First, it reduced markets for traditional food crops, a trend which was reinforced by government policies which deliberately set low prices for such crops. Second, it created an expanding livestock market, which the southern rainforest farmers found difficult to supply, due to pests and diseases such as the tsetse fly. These trends encouraged some savanna grain farmers to turn to animal-rearing, so that they no longer welcomed the annual in-migration of the Sahelian herders.

The farming changes in the savanna zone closed an important safety valve for the Sahelian herders, whose growth in numbers was already exacerbating a deteriorating environment. The desert margin is believed to have been 300–400 kilometres further north some 25 000 years ago. The ending of the last ice-age caused a strengthening of high pressure zones over the Saharan region and created a stronger obstacle to the northward shift of the rain-bearing ITCZ (Inter-Tropical Convergence Zone). There is also evidence this century of a 30-year cycle, in which 5 years of drought are preceded by relatively abundant rainfall – during which herd sizes are expanded. The recorded median years of the dry periods have been 1913, 1942 and 1971, when rainfall fell short of the norm by 41 per cent, 21 per cent and 30 per cent, respectively. Historically, the droughts reduced both the size of the herds and also the human population.

The diffusion of western medicine, which is one of the first changes associated with economic development (Gould 1970), caused the birth rate in the Sahelian region to shoot up to 45–50/1000, while the death rate fell to 20–25/1000, so that the population had a doubling rate of around 30 years. This rate of increase could be accommodated during the wetter years by expanding the herds, but little attempt was made to improve the quality of the animals, pasture or water supply. Herders rejected arguments that fewer animals would improve the quality and gross value of their livestock. Pearce *et al.* (1990) note a ratchet effect in herd expansion under these conditions: during wetter years herders draw down their stock, but at a slower pace than the overall rate of increase. During drier years they reduce the rate of stock withdrawal and seek to hold numbers steady.

In the absence of either alternative grazing or alternative work opportunities in the cities, the natural pasture deteriorates. First the perennial grasses, which grow up to 2 metres tall in the rainy season and put down deep roots of a similar size, die out. Those roots tap the underground water-table, but as grazing intensifies they become shallower and the perennials are replaced by coarse annual grasses which in turn succumb and are replaced by leguminous plants which dry up quickly and have less ability to bind the soil. The hooves of the cattle loosen the soil and make it more susceptible to both wind and rain erosion. Meanwhile, the environmental deterioration is aggravated by worsening fuel-wood shortages caused by demand for charcoal from the regional towns and also from Middle Eastern countries. This led to the removal of bushes without restocking (Hardin 1977).

Periodic droughts accelerate the deterioration. During the 1969–74 Sahelian drought, there was a 30 per cent reduction overall from the expected rainfall. Concentration of the herds around the shrinking water holes led to massive overgrazing and a breakdown of herd rotation. Many pastoralists lost their herds and were forced to squat on the outskirts of the cities. Some 250 000 people are estimated to have died of starvation, an outcome which in part reflected the difficulty of distributing food in a region some 1000 kilometres from the coast with few improved local roads.

In 1977 the Club des Amis du Sahel proposed five main measures to double livestock herds and cereal production by the year 2000. One measure was the breeding of cereal varieties which would mature in a shorter rainy season. A second change involved the use of desert-tolerant crops, such as jojoba oil and the kad tree, to help stabilize the soil and also to provide new sources of income. Meanwhile, more careful use of the available water could be achieved by 'drip' application to crops. The construction of more dams and bore holes would improve the capture and storage of rainfall, both in the wadis (dry valleys) and in the underground water-table. Finally, the Club des Amis du Sahel proposed a restocking of the rangelands with new seeds.

Such proposals merely evoke scepticism, because prior to 1969, the drilling of boreholes and the vaccination of cattle had merely led to overloading of the grazing lands, intensifying the adjustment pressures when the rains failed. Chad lost one-third of its livestock and Niger two-thirds during the 1969–74 drought. Further east, in Somalia, transmigration was attempted with some 100 000 nomadic refugees, who were flown to six new settlements in the south of country. They were helped by UNICEF and the World Bank to develop 15 000 hectares of farmland in an effort to encourage the adoption of a sedentary existence. The initial response was good, with only a relatively small number wishing to return to the nomadic life. But the dry-farm zone environment also is vulnerable to growing population, as the next section shows.

The savanna dry-farm zone

The adverse environmental consequences of increased population pressure for the savanna zone of dry-farming may be illustrated with reference to Ethiopia. There, the dry tropical forest has shrunk at a rapid rate, to occupy barely 4 per cent of the land area. As a consequence, soil erosion has intensified. A 1986 FAO study estimated that over 1900 million tonnes of soil were being lost annually from the Ethiopian highlands. If that trend persists, then by the year 2010 some 38 000 square kilometres of the region would be eroded down to bare rock and a further 60 000 square kilometres would have a soil depth of only 10 centimetres, the minimum depth required to support cropping (FAO 1986).

As population pressure mounts, a five-stage deterioration in the dry-farm environment occurs as the acacia tree cover is thinned and the nutrient cycle is degraded (Newcombe 1989). The thinning occurs because wood is the principal source of fuel in Ethiopia, in both rural areas and urban areas. Without it, the country would need to import an estimated 1 million tonnes of kerosene annually, and it lacks the foreign exchange to do so.

The first stage of environmental deterioration is triggered when the timber harvest exceeds the rate of regrowth. In effect the timber ceases to be a flow (or renewable) resource, and becomes a fund (or finite) resource. In the second stage the timber is increasingly sold outside the region to urban markets and more distant localities. The peasants must resort to the collection of dung and straw as a substitute for their own fuel needs. This means that in addition to the loss of nutrients (and rain protection) from the acacia trees, the topsoil is also increasingly deprived of crop residues and dung. The topsoil therefore has a lower proportion of organic matter, weaker structure and greater risk of erosion. In stage three the entire tree cover is removed and both crop residue and dung are collected and sold, in part, to urban markets.

The negative feedback loop then gathers pace: cereal yields in the dry-farm zone fall, but the area harvested also contracts because of declining livestock numbers. In stage four of the cycle, all the crop residue goes to feed the animals while all the dung is sold to the urban markets, where fuel shortages are reflected in rising prices. By this stage, both the arable and grazing lands are now bare much of the year, with deleterious effects on nutrient depletion and soil erosion. The system becomes vulnerable to shocks, such as prolonged drought, so that starvation and the devastation of animal numbers become likely. This causes the peasant farmers to abandon their land in search of sustenance in the cities, marking the fifth and final stage of the cycle. But the surge of rural–urban migration merely aggravates the shortage of fuel in the cities, pushing prices up even further and drawing more distant farming zones into the downward spiral.

By the final stages of the cycle, the value of dung as fuel greatly exceeds its value as a source of fertilizer. These conditions increase the incentive to sell dung for cash rather than use it for fertilizer to produce more grain.

One solution to the cycle would be rural afforestation, which, according to Newcombe (1989) could both produce a higher quality fuel at one-quarter the price of dung *and* improve soil protection and soil nutrient content. But the unhappy outcome of a degraded rural environment and a collapsing farming system is intensified where, as in Ethiopia and much of sub-Saharan Africa, government policies depress food crop prices and fail to earn sufficient foreign exchange to allow the substitution of imported kerosene for biomass fuels.

Hard technology solutions

The environments discussed above are not inherently poor, and some observers have suggested that modern technology can ameliorate many of

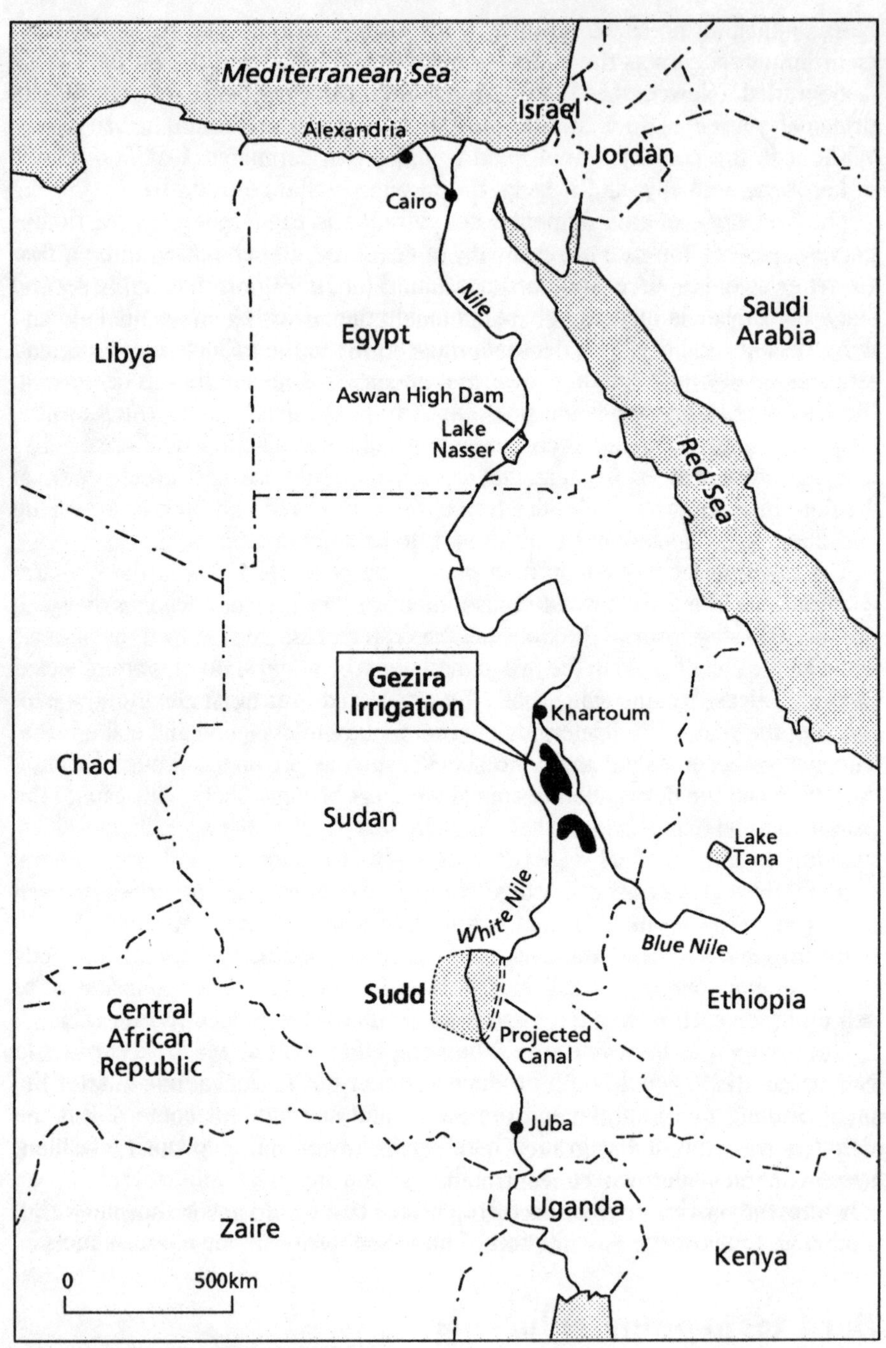

Fig. 2.4 *The Nile Basin*

the problems. For example, Hopper (1976) estimates that modern technology could produce very high levels of output in the Sahelian zone. This is because the Sahel has very high rates of insolation – much higher than the cloudy rainforest zone, for example. This makes areas like the southern Sudan, which also possess relatively fertile and water-conserving soil, potentially highly productive if a reliable supply of water can be provided.

Hopper suggests that a substantial part of the Sudd region of southern Sudan could be transformed from a vast waterlogged swamp straddling the Nile (Fig. 2.4) into a major global grain-producing region. It would require appropriate drainage and water management, along with a substantial expansion of irrigated agriculture. New high-yield crop varieties and modern scientific farming methods could, from three crops per year, produce 20 tonnes of grain/hectare. Hopper calculated that the total production from this scheme alone could reach 1 billion tonnes of grain annually, a figure close to the entire world grain output in 1970. In line with such a projection, *The Economist* (1993a) estimated that Africa contains half the world's unused farmland which, if it attained the best yields obtained elsewhere on similar soils, could grow at least one hundred times more food than presently.

But evidence from elsewhere in Africa suggests that the optimism of the 1950s and 1960s concerning the ability of large technological schemes to solve African problems is misplaced. Experience in the Nile Valley north of the Sudd suggests that large-scale solutions often bring unforeseen problems, including difficulties with maintenance, so that they may fail to resolve the fundamental problem.

Hard-tech fixes in the lower Nile Valley

Rationale for the Aswan High Dam

Relative to its resources, Egypt was already an over-populated country at the beginning of the nineteenth century, whose development required industrialization, so that manufactured exports could pay for imported food. The population of Egypt is estimated to have been around 2.5 million at that time, with three-quarters living in the Delta region where much of the 1.5 million hectares of farmland was concentrated.

In fact, the basis for a manufacturing sector that would absorb surplus rural labour was already emerging with the expansion of cotton production. Further growth was prevented, however, by the intervention of European colonial powers, because Egyptian textile production was in direct competition with the mills in north-west Europe. Consequently, by 1950 the economy of Egypt was still dependent on farming, although the population by then had grown to almost 20 million and supported itself at a low per capita income on 2.2 million hectares of cultivated land.

A successful revolution led by Nasser in 1952 set about improving the rural economy and industrializing the country through the construction of a large multi-purpose dam, the Aswan High Dam. It was built with Soviet aid at a cost of around $1 billion and is the epitome of the technological fix. The Aswan High Dam had five key functions, namely:

- reduction of the flood risk;
- extension of the irrigated area;
- improved drainage in the Delta region;
- the creation of a fishery in the lake created by the dam;
- the generation of 10 000 MW of hydroelectric power.

The flow of the Nile is highly seasonal, reflecting conditions along its two principal tributaries, the White and the Blue Nile. The White Nile rises on the East African plateau, and its flow at Khartoum averages 110 million cubic feet of water daily from October to November, compared with only 20 million in June. The seasonal variation is much larger, however, on the Blue Nile which rises in Ethiopia. The flow at Khartoum peaks at 650 million cubic feet daily in September, reflecting the impact of the monsoon in the eastern horn of Africa, and falls to a mere 10 million cubic feet daily in April.

The Nile also experiences long-term annual variation in its flow. It had a mean annual discharge of 84 billion cubic metres in the century prior to the construction of the dam. The standard deviation was 20 billion cubic metres, but the recorded extremes were a low of 42 billion cubic metres, which occurred in 1912, and a high of 150 billion cubic metres, in 1887. The regulation of the river flow by the Aswan High Dam therefore promised to remove the threat of flood and also to greatly reduce the risk of drought in the Lower Nile Valley.

The Aswan High Dam would also extend the irrigated area by bringing water to an additional 0.4 million hectares and allowing the conversion of a further 0.3 million hectares from basin irrigation (which uses the natural autumn-flood water to produce a single crop for harvest in April–June). The conversion to perennial irrigation would permit up to three crops to be taken annually. In addition, modifications to the river flow would lower the water-table in the Lower Delta region and improve drainage there. In the lake behind the dam, fish catches were expected to exceed 50 000 tonnes annually. Finally, the hydroelectricity generated by the dam would permit both an extension of electrification and the growth of energy-intensive industry such as aluminium smelting.

The disappointing outcome

The Aswan High Dam was unveiled in 1971 to a wave of criticism from environmentalists. Fears were expressed about the safety of the dam, because its considerable weight (the dam itself is 70 metres high and 1 kilometre wide at its base) has triggered earthquakes, which might inflict structural damage. Meanwhile, the loss of water from the lake (which extends for 500 kilometres upstream from the dam), at 8 per cent of the total normal storage capacity of 150 billion cubic metres of water, was higher than expected. In addition to water losses due to evaporation from high temperatures and high winds on the surface of the lake, leakage into adjacent rocks was also greater than expected.

Lake Nasser yielded less fish than expected, failing even to offset the loss of fish at the mouth of the Nile caused by the loss of silt and plankton. Nor were the agricultural gains as large as expected. The peasant farmers were inadequately trained, and drew too much water into their canals. Inadequate attention was given to drainage and this caused extensive waterlogging and salinity, rectification of which may require expenditure in excess of $1 billion (as much as the total pre-1971 Aswan investment).

Of the 400 000 hectares of extra land brought under irrigation during 1960–80, only two-thirds was under crops, and half of that failed to cover its costs, due mainly to the poor quality of the soil. Meanwhile, urban expansion absorbed 10 000 hectares annually as the population grew inexorably. Having doubled to 19 million in the 50 years to 1947, the Egyptian population then doubled again to 36 million in 1976 and is projected to reach 65 million by the year 2000 (Beaumont *et al.* 1988). Domestic food demand had doubled between 1960 and 1980 and more than half of all food consumed was imported by the 1980s.

Far from slowing, the growth rate of the Egyptian population accelerated after the 1970s to 2.7 per cent/year. Nasser's successor, Anwar Sadat, was unsympathetic to birth control and argued that raising income would achieve a deceleration in population growth. In fact, with a ratio of one doctor and one clinic to 5000 people, Egypt was in a relatively favourable position compared with many developing countries to pursue a policy of birth control. In the absence of strong government backing, however, population growth had outstripped the disappointing resource expansion of the Aswan High Dam by the 1980s.

The technological fix was considered by some observers to have merely postponed the real problems which the country faced, namely over-rapid population growth and the inefficient use of existing human and physical resources. In addition, inclusion of the full environmental and social impacts of large dams in the feasibility studies has sharply reduced the attractions of schemes involving large dams in both arid and humid tropical zones alike (Mikesell 1992). Nevertheless, the Egyptian government persisted with its ambitious plans for land colonization.

Proposed expansion of the cultivable area by 2000

The Egyptian government announced plans in 1980 to add a further 1 million hectares of irrigable land at an estimated cost of $12 billion, or some

$10 000/hectare. As shown below, this was a relatively expensive option. Moreover, the plan may not be feasible: each hectare requires 40 000 cubic metres of irrigation water per annum. The annual 40 billion cubic metre requirement would represent a 60 per cent increase on the country's existing usage rate. The Egyptian government believed it could secure almost a quarter of the additional water resources through measures to reduce evaporation rates in the Sudd region of southern Sudan (Fig. 2.4). A slightly larger saving could be secured through more efficient water use in existing irrigated areas, with the introduction of tile drains and better education of farmers. But even if these objectives were achieved, the country would still have barely half the extra water resources it needed.

The World Bank criticized the extension of the cultivated acreage as a misuse of capital, because it increased output in a manner which was several times more expensive than achieving a similar rise in farm output through the intensification of production on existing land. In other words, a superior strategy would have been to make better use of the existing water system through a combination of policy reforms. The latter include land reform and the removal of disincentives to farmers.

The World Bank considered further land reform a prerequisite for raising the growth rate in agriculture to the 4 per cent per year level needed to get ahead of population growth. The land reform of the 1952 revolution had proved only an interim solution. By 1978, despite a ceiling of 40 hectares per family, some 95 per cent of the country's 3.5 million peasant farmers held less than 1 hectare of land each, whereas three-fifths of the land was held in farms of over 20 hectares. Further land redistribution would be ineffective, however, in the absence of measures to boost productivity on the redistributed fields.

Land reform also needed to be accompanied by an extension of cooperatives (to capture the economies of scale) and of the farm advisory service. In addition, credit for small farmers needed improving, so that more productive methods could be introduced. One-quarter of all farmland was under clover for livestock fodder (an area greater than that in wheat and rice combined) so that land could be freed for the production of non-fodder crops if the buffalo waterwheels used in irrigation were replaced by diesel pumps.

But, as elsewhere in Africa, a fundamental requirement for agricultural progress was reform of farm prices. Egypt set grain prices low in order to reduce the cost of living of the poor. This policy, however, reduced the incentive to farmers to grow food, and they preferred instead to grow export crops such as cotton, fruit and meat products. By the early-1980s, a massive 15 per cent of GDP was being expended on food subsidies, basically buying imported wheat at world prices and selling it on the domestic market at low prices. This policy also disadvantaged industry, because food imports received priority over the purchase of imported industrial inputs in the allocation of foreign exchange.

The World Bank recommended phasing out the food subsidies by freezing the absolute size of the subsidies so that they would decline in real terms over time, allowing the market to determine the prices of food and other crops. Meanwhile, a diversification of both industrial and agricultural exports was required, because by the mid-1980s an expansion of oil production was pushing the country towards the status of a mineral economy, with all the risks of over-dependence on a single commodity which that entails (see Chapter 9). The high level of oil dependence is exacerbated in the case of Egypt by a heavy reliance on the remittances of guest workers employed in neighbouring oil-exporting countries like Saudi Arabia.

Decay in the Sudanese Gezira scheme

An earlier large irrigation scheme further up the Nile in Sudan illustrates the problems which African governments may have in maintaining technologically elaborate systems. The Gezira scheme is a large irrigated area which was established under British colonial rule in the 1920s to supply cotton to Lancashire. It lies on clay soils south of Khartoum, between the White and Blue Niles, and occupies around 800 000 hectares (see Fig. 2.4). In 1970 it provided one-third of Sudan's GDP, but it was already in decline (Pollard 1981).

One important reason for the decline of the Gezira scheme is the pricing policy adopted by the Sudanese agricultural marketing board from the early-1950s, which squeezed incentives to the farmers. A second problem is insufficient attention to maintenance, which led to a deterioration in the water system infrastructure. The productivity of the agricultural system is running down and requires an overall policy change in favour of the countryside.

As in Egypt, the government of the Sudan needs to shift scarce resources away from the towns and towards the countryside, because the towns require much greater infrastructure per person and yet many of the activities located there, notably manufacturing, are inefficient and require onerous subsidies from the rural sector. The government also needs to raise incentives(and therefore incomes) for farmers by reforming food prices in order to bring them into line with world prices. Meanwhile, farm productivity needs to be increased in the Sudan on *existing* land, before turning to expensive new water management schemes like the risky Sudd drainage project.

Conclusion

The disappointing post-independence development experience of most African countries is rooted in inadequate domestic economic policies in the face of a generally deteriorating set of export prices. The political economy of the adoption of such policies reflects the neglect of rural interests in favour of the rapid expansion of (frequently inefficient) urban-based industrial and administrative activity. The policies have discriminated against small farmers in particular, despite the fact that they and their families constitute a majority of the population.

Prolonged economic stagnation, combined with growing population, has exacerbated environmental problems, as the examples from the Sahel and Ethiopia illustrate. But despite the undoubted farming potential of the arid zones, hopes have faded for the success of large-scale technological fixes like the dams and associated irrigation schemes in the Nile valley.

The experience of the Gezira scheme in the Sudan reinforces that of Egypt. Where land is in short supply, it is more efficient to improve the productivity of existing land resources, before embarking on large-scale, capital-intensive schemes of land colonization. But, as the next chapter shows, much of Africa south of the Sahel is not short of land, and it requires its own green revolution in which small-scale schemes are likely to be favoured over large-scale ones and less emphasis is placed initially on boosting land productivity.

Achieving the appropriate relationship between population, technology and land use calls for new macroeconomic and local policies, if Africa is to have a second chance after its false start. These issues are discussed in Chapter 3, which begins with a systematic examination of the relationship between population and economic development, before exploring the links between technology and land use under increasing population pressure.

3

Population, technology and land: a soft-tech greening of Africa

Population and agricultural technology

Much of Africa has experienced rapid population growth with falling real incomes through the 1970s and 1980s, a process which is increasingly degrading the rural environment. Improvement requires the adoption of a rural-based development strategy which steadily raises farmer productivity without damaging the environment.The more optimistic observers, like Hopper (1976), have argued that the application of western technology can substantially increase production above present levels. But African experience with such schemes along the Nile Valley and elsewhere indicates that such technology is no panacea.

Yet the technological optimist's view persists. For example, Revelle (1984) draws on FAO figures to suggest that with a potential cultivable area of 1.1 billion hectares (excluding South Africa), the continent could support a population of over 10 billion people, a figure well above that at which the African population is projected to stabilize. Revelle's data assume an average daily per capita intake of 2350 calories and crop yields with intermediate farm technology averaging the equivalent of 3 tonnes of cereals per hectare. He estimates that the land available comprises 110 million triple-cropped irrigated hectares in the humid tropics and 290 million irrigated hectares plus 490 million rain-fed hectares outside the humid tropics.

But even intermediate farm technology is inappropriate for much of sub-Saharan Africa in the 1990s. In any case, intermediate technology would be difficult to adopt because of the policy and institutional failures identified in the previous chapter. A different approach is therefore required in order to accommodate rising population in an African context.

The present chapter explores the relationship between population, technology and land use in more depth, beginning with a review of the controversy over the role of population growth as a catalyst for change. Some observers consider population growth to be a cruel, but necessary, catalyst of change; others regard it as a brake on economic progress. Some support for the first view is found in the adjustment in parts of sub-Saharan Africa to mounting population pressure, but persistent rapid population growth does appear likely to depress long-term living standards. Subsequent sections of the chapter explore the need for more intensive land use in different parts of Africa, outlining the adjustments required to achieve a greening of farming which is appropriate to the land availability and institutional capacity of the sub-Saharan African region.

Population pressures

A disappointing rate of economic growth is especially problematic for low-income countries like those of much of Africa, because of the stage they have reached in the demographic transition. The latter is the five-stage process whereby countries shift from a relatively stable pattern of high birth rates and high death rates, to a new equilibrium of low birth rates and low death rates typical of the industrial countries. The three intervening stages are:

- high birth rate and falling death rate, as western medicine diffuses into hitherto isolated communities – a process traced for Tanzania by Gould (1970);
- high birth rate and a low death rate;
- falling birth rate.

The third stage of the demographic transition model gives the highest rate of population growth, with the birth rate around 45/1000 and the death rate dropping to 12/1000. This is the stage that much of sub-Saharan Africa had reached in 1990: the birth rate was estimated at 46/1000 and the death rate at 16/1000, giving a population growth rate of 3 per cent. This growth rate compares with 2.1 per cent in South Asia, 2.0 per cent in Latin America and the Caribbean, and 1.6 per cent in East Asia (World Bank 1992a). The long-term implications of these differing population growth rates for the principal developing regions are summarized in Fig. 3.1. The population of the African continent is projected to triple to almost 1.5 billion over the period 1980–2020, while that of Latin America only doubles to a level half that size. In the longer term, the World Bank projects that the population of Africa will rise to 3 billion, or almost one-quarter of the global total, before stabilizing around 2075 (*The Economist* 1993a).

The demographic transition model assumes that urbanization plays a key role in curbing population increase. This is because the city provides more opportunities for all members to contribute to the family income, so that even if the breadwinner achieves only a modest increase in pay, the family as a whole experiences a higher income. Such increased income reduces the incentive to have more children in order to provide for the parents' old age. Meanwhile, the crowding associated with urban residence encourages limits

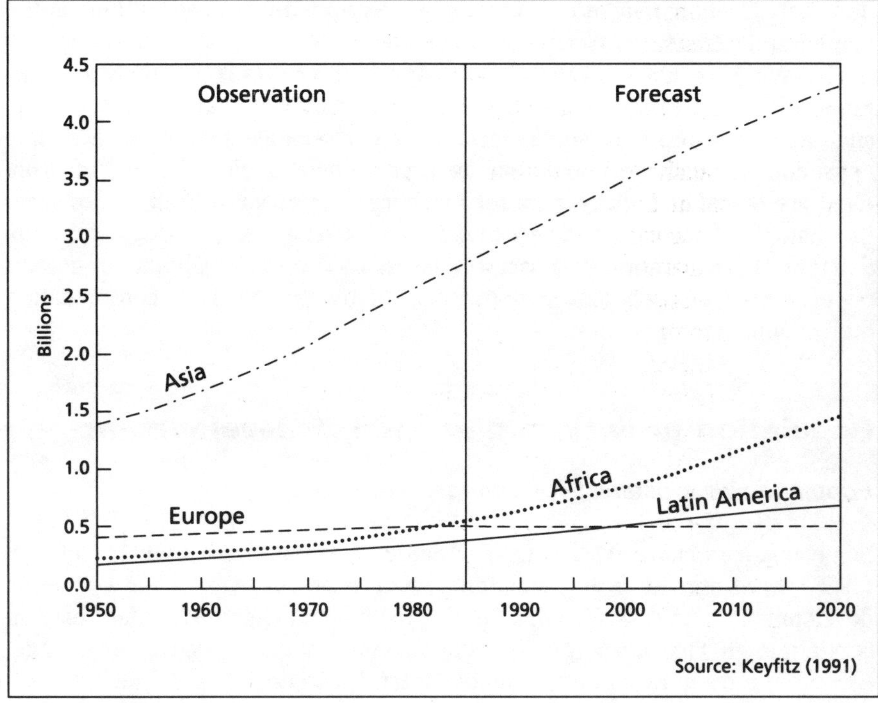

Fig. 3.1 *Population growth projections, 1950–2020*

on the size of the family, while the greater anonymity of urban life gives greater freedom from the social pressures of the village to conform to traditional views concerning religion, the role of women and the size of families.

The demographic transition model exhibits some discrepancies, however. For example, although as noted above, low-income sub-Saharan Africa fits the overall model, Zambia has a population growth rate of 3.4 per cent which is above the regional average, despite the fact that its level of urbanization is twice the regional level. Similarly, the figures for low-income South Asia (2.1 per cent) and China (1.5 per cent) are lower than the transition model would suggest, while highly urbanized Latin America has higher rates than the model would predict.

It is clear that the demographic transition model takes insufficient account of the important mediating role played by cultural factors. Societies where the status of women is not inferior are quickest to seize the opportunity to reduce family size (as in Communist China and Hindu India for example, in contrast to Catholic Brazil and Muslim Egypt). Moreover, some researchers like Ward and Dubos (1972) argue that the potential of the green revolution in agriculture to raise living standards in the rural areas of Asia, means that high urbanization may no longer be a prerequisite for completion of the demographic transition to population stability.

This suggests that a deceleration in population growth is not likely to occur humanely in an economically neglected and male-dominated rural society, such as that of much of sub-Saharan Africa. The consequences of a rapidly expanding low-income population on a potentially fragile physical environment are therefore rightly a matter for increasing concern. Before exploring the nature of an appropriate rural-driven development strategy for sub-Saharan Africa, it is first necessary to assess the impact of population growth on economic development and the role of technology in accommodating higher population densities.

Population growth and economic development

Population as a catalyst for change

Some scholars, notably the Latin American economist Albert Hirschman (1958), have argued that population growth may positively affect economic development. The case for population growth as a positive stimulus rests on two arguments concerning, first, the capture of economies of scale and, second, the triggering of a technology-learning process.

The capture of scale economies arises from the fact that much economic activity in western economies has a U-shaped or L-shaped long-run cost curve. This implies that the average cost of manufacturing goods or providing services initially falls markedly with increased production/provision, and then either stabilizes (giving a so-called 'L-shaped' curve) or rises (the 'U-shaped' curve). Many developing countries have national markets which are too small to support an industrial plant of minimum viable size. For example, the national market of all but the largest developing countries has less demand for goods than Kansas City or the West Midlands. Similarly, many rural communities in low-income developing countries are too dispersed and poor for there to be sufficient concentrations of people/money to support the efficient provision of schools, roads and clinics (Johnson 1970).

Under these conditions of small domestic markets and dispersed rural populations, an increase in population contributes to the more efficient use of resources by propelling national and/or regional populations beyond the threshold required for the efficient provision of manufactured goods and services. Put another way, as the population grows, more and more goods and services can be supplied at an efficient scale and therefore at a lower cost per capita.

The second expected developmental benefit from rapid population growth concerns the technology learning curve. Historically, population pressure has stimulated innovation by exerting a downward pull on per capita living standards, which demands technical and social change in order to restore former levels of well-being, creating the conditions under which traditional societies break with historical practices. Efforts to offset falling incomes teach the community a greater mastery of the environment and also lower the tolerance for rigid social stratification. Another way of conceptualizing this is to say that the quality of human resources, or social capital, is advanced by the learning process which the prevention of income loss entails.

In line with these arguments, Hirschman sees population growth as a critical – if cruel – catalyst for change, which operates through a three-stage sequence. First, the learning process begins as a result of efforts to counter the depressing effect on living standards of larger numbers of people. Second, the scale benefits accrue as innovations run ahead of population growth, giving more efficient use of national economic and physical resources. Finally, population growth slows when sustained growth has become built into the economy.

Such views are consistent with Hirschman's broader thesis of 'unbalanced growth'. That thesis was developed in the 1950s, and it rests on the premise that the critical shortage in developing countries is not capital (as was widely thought at the time), but decision-making skills. Consequently, Hirschman argued that rather than delay investments for fear of error, developing countries should espouse unplanned growth because it creates self-rectifying tensions. For example, if the construction of a new power station results in a greater than expected increase in traffic flows, the bottle-neck will automatically call forth the required investment to ease the traffic flow. Similarly, the growth in population in rural areas is another such self-correcting tension which Boserup (1965) argues acts as a catalyst for change.

Population as a constraint on development

Other scholars such as Coale (1964) have long argued that rapid population growth acts as a drag on economic development. Coale illustrates his case with reference to simple simulations of economic growth in two countries, which he describes as High Fertility (HF) and Low Fertility (LF), and which have similar initial population/resource endowments. The HF model assumes a sustained high level of population growth, such as that occurring in much of sub-Saharan Africa in the 1990s. In contrast, the LF model assumes a 50 per cent reduction in fertility over a 25-year period. Coale assumes that the percentage national income which needs to be reinvested annually for sustained growth is more than 10 per cent. He then spells out the implications of the two models for incomes, investment and welfare.

With reference to income, after 25 years the workforce of the HF model has increased by only 4 per cent more than the LF model, but the HF country already has 50 per cent more dependents (pregnant women and young children) to support. The dependency ratio (which measures those able to work against those who cannot) is 94/100 for the HF model and only 65/100 for the LF country. This means not only that each worker in the HF model must support more mouths, but that the HF country must divert a larger percentage of investment than the LF country to primary education and house construction, areas of expenditure which have a very long payback period. The HF country must also spend more on immediate consumption, while the LF model has a greater capacity to save and reinvest. As a result, the LF country emerges with higher total economic output and *much* higher income per capita than the HF country after only 25 years.

Turning to investment, if it is assumed that the value of all the national capital stock is three times the national income (an empirical rule of thumb), then in order to prevent a fall in labour productivity (let alone achieve a steady rise), the ratio of investment growth to labour force growth must be 3 to 1. This means that, with 3 per cent annual growth in the labour force, investment must rise by 9 per cent each year just to maintain existing levels of productivity. With only a 1 per cent growth in the labour force (as with the LF model) a modest 3 per cent rise in investment prevents a decline in productivity, while an additional 6 per cent increase in investment raises productivity significantly. As a result, the LF country has a less onerous path to higher productivity and higher per capita income.

Finally, with regard to welfare, not only can the LF country increase productivity faster than the HF country, it can also consume more while doing so. The LF population is therefore better nourished and better cared for than that of the HF country. The population of the LF country would also be less apathetic and therefore more productive than the HF population. Thus, within 25 years – a single generation – the LF country would display a marked improvement over the HF country. To appreciate the full extent of the environmental implications, if the two models ran for four generations, the population of the HF country would be 20 times the size of the base year population, compared with a population only 4 times larger for the LF model.

There seems little doubt that in those sub-Saharan African countries where economic management has been severely flawed, any 'cruel catalytic' effect of population growth is taking an uncomfortably long time to operate. Meanwhile, any benefits arising from the capture of the cost savings from the economies of scale and progress along the technology learning curve have yet to compensate for the depression of living standards. Hirschman had little to say about the consequences of environmental degradation while the learning curve takes effect. It is therefore necessary to examine more closely the prospects for coping with population pressure in the specific context of low-income Africa.

Technology and land use: stages of agricultural intensification

Traditional farming systems: swidden

Five stages may be recognized in the transition from traditional farming systems to western scientific agriculture. The five stages are characterized by

Table 3.1 *Intensification of tropical cropping systems*

Systems	Farming intensity[a]	Population density (Persons/ sq. km)	Tools used
Forest fallow	0–10	0.4	Axe, machete, digging stick,
Bush fallow	10–40	4–64	Axe, machete, digging stick, hoe
Short fallow	40–80	16–64	Hoes, animal traction
One crop/annum	80–120	64–256	Animal traction

Source: Pingali and Binswanger (1988)
[a]Index = number of years of cultivation x 100/number of years of cultivation – number of years of fallow

increasing intensity of land use, and they are shifting cultivation, rain-fed perennial farming, traditional irrigated farming, quick-return scientific farming (the green revolution) and industrial scientific farming.

Based upon research in the Sahelian zone (Shepherd 1991), in its basic form the fields in the swidden system are rotated. The fields also display variations in intensity with distance from the villages. Intensive vegetable production is practised in kitchen gardens adjacent to the household compound, while a short fallow zone of permanent fields may encircle the village with a long fallow zone surrounding that. In some Sahelian systems the villages themselves appear to rotate every 120–150 years.

Swidden systems are well-adapted to the environment (Dickinson 1972): the inter-cropping of plants on small plots maximizes the use of nutrients at varying depths, because of the differing crop root-lengths. A second benefit of this crop symbiosis arises from the fact that appropriate crop combinations offer soil canopy and leaf shade to those crops which need it. Moreover, the mix of crops insures the subsistent farmer against loss from extremes of weather, because the range of plants means that some will survive whether conditions are unusually wet or dry. The mixture of crops also gives a varied diet and distributes food production over the cropping season.

The main criticism of swidden agriculture is its relatively low yield per hectare per year, and therefore its low population carrying capacity. Although, as Table 3.1 shows, land use can be intensified through a gradual shortening of the fallow period (from a 10-year forest fallow to a shorter 5-year bush fallow, and finally down to a 1- or 2-year grass fallow), such a process is not without environmental and social risks.

For example, a study of the Tiv in south-eastern Nigeria (Vermeer 1970) revealed a high risk of laterization (which transforms the soil into a hard pan) as the fallow period is reduced. The Tiv also experienced social problems as a result of the changes in the principal crops which shorter fallowing required. More specifically, a reduction in yam yields (traditionally grown in the first year of the 2-year field rotation cycle) resulted in a 'hungry' season before the second crop ripened. The yam was therefore replaced by quick-ripening peanuts or sweet potatoes as well as sesame or millet in the same season. This, however, put strains on the social system because yam, as the 'king' crop, had determined customs, status and ceremonies – as described so vividly in Achebe's (1958) novel, *Things Fall Apart*. Interesting links between crops and society are also recorded from New Guinea by Rappaport (1971), in an analysis of energy flows in a swidden system.

Nevertheless, swidden systems can be successfully transformed into permanent cultivation without environmental degradation or social failure (Clayton 1964). Table 3.2 traces changes in farm inputs with the evolution of rain-fed farming systems from forest fallow to perennial cultivation. In addition to shortening the fallow period, physical modifications to the land are also made, a first response often being to improve the land by removing stones and levelling plots. Dykes may be built to improve drainage, and irrigation canals may eventually be constructed. Such changes effectively make more land available, but when land is in plentiful supply they are not worth the effort expended.

Other changes affect the tools used and also the application of nutrients to raise yields. While the labour inputs required by a farming system are modest (as with the swidden farming system), simple hoes are often adequate. Hoes will only be replaced, first by animal-drawn ploughs (assuming the absence of tsetse fly) and then tractors, when the economic returns justify it. Finally, chemical fertilizer is unlikely to be remunerative until the switch to perennial cultivation has been made (Table 3.2).

The intensification of agriculture is likely to entail a change in local soil preferences, as illustrated in Table 3.3. The soils on upper slopes are initially preferred because they are relatively light and easy to work by hand, so that tillage requirements are minimized. Moving down the slope, the clay content of the soil increases; this raises the weight of the soil, demanding greater work effort. The lower lands on the slope are the most difficult to work, and they are often left uncultivated when population densities are light. Yet these same bottom lands have a higher moisture content and are also less prone to erosion: they therefore respond more productively to investments of labour and to land improvements (such as drainage). Consequently, as population densities rise and the relative scarcity of land and labour reverses (so that labour becomes more plentiful), there is a reorientation of farming away from the upper slopes, towards the more productive bottom lands.

There is encouraging evidence that areas of modest natural potential, which historically were regarded as environmental disasters and required food imports and famine relief, can recover to support higher levels of population, with minimal environmental damage. Tiffin *et al.* (1994) examined land-use change in the Machakos district of Kenya between 1930 and 1990. In the

Table 3.2 *Operations in different traditional farming systems*

Operation or situation	Forest fallow system	Bush fallow system	Short fallow system	Annual cultivation system	Multiple cropping system
Land clearing	Fire	Fire	None	None	None
Land preparation and planting	No land preparation; use of digging stick to plant roots and sow seeds	Use of hoe and digging stick to loosen soil	Plough	Animal-drawn plough and tractor	Animal-drawn plough and tractor
Fertilization	Ash, perhaps household refuse for garden plots	Ash, sometimes chitimene techniques[a], household refuse for garden plots	Manure, sometimes human waste, sometimes composting	Manure, sometimes human waste, composting, cultivation of green manure crops, chemical fertilizers	Manure, sometimes human waste, composting, cultivation of green manure crops, chemical fertilizers
Weeding	Minimal	Required as the length of fallow decreases	Intensive weeding required	Intensive weeding required	Intensive weeding required
Use of animals	None	Animal-drawn plough begins to appear as length of fallow decreases	Ploughing, transport interculture	Ploughing, transport interculture, post-harvest tasks, and irrigation	Ploughing, transport interculture, post-harvest tasks, and irrigation
Seasonality of demand for labour	Minimal	Weeding	Land preparation weeding and harvesting	Land preparation, weeding and harvesting	Acute peak in demand around land preparation, harvest and post-harvest tasks
Supply of fodder	None	Emergence of grazing land	Abundant open grazing	Open grazing restricted to marginal lands and stubble grazing	Intensive fodder management and production of fodder crops

Source: Binswanger and Pengali (1988)

[a] To augment the ashes from the bush cover, branches are cut from surrounding trees, carried to the plot of land to be cultivated and burned to provide extra nutrients for the soil.

Table 3.3 *Farming intensity, agro-climates and soil preferences*

	Farming intensity		
Agro-climates	Forest + bush fallow	Grass fallow	Permanent cultivation
Arid	Lower slopes and depressions only	Lower slopes and depressions only	Lower slopes and depressions only
Semi-arid	Mid-slopes	Plus lower slopes	Plus depressions
Sub-humid	Upper slopes	Plus middle and lower slopes	Plus depressions
Humid	Upper slopes	Plus middle and lower slopes	Plus depressions

Source: Pingali and Binswanger (1988)

1930s, most farmers reared cattle and grew some grain and pulses in a landscape scarred by soil erosion.

An intensification occurred in the Machakos region's carrying capacity, linked to the incorporation of capital and ideas brought from outside the region by returning migrants with indigenous knowledge and skills. The creation of non-farm employment and access to external markets for cash crops were also important elements in diversifying and increasing sources of income, a point also stressed by Goldman (1993) in a comparative analysis of rural change in Kenya. Between 1930 and 1990, population within the district rose five-fold, yields rose ten-fold and output increased fifteen-fold as even erstwhile marginal land was brought into production. Land reform encouraged investment in soil and water conservation, as well as in the production of higher-value cash crops. Some 70 per cent of the cropped land was gradually converted to terraces, with significant gains in soil conservation and land productivity. But it is important to note also that until the late-1980s, economic management in Kenya was relatively sound (Killick 1983).

Traditional farming systems: sawah

The shift to permanent rain-fed cultivation may raise yields of cereals per hectare to around one tonne (Hopper 1976), but further increases require substantial modifications of the farming system, including changes in both technology and land use configurations. The third stage of land-use intensification is already heralded in the droughty area examined by Tiffin *et al.* (1994) and involves environmental modification through the construction of traditional irrigation systems like the tanks of South Asia (where water is stored in hill-side tanks and used to irrigate valley land and depressions) and the terraced sawah system of South-East Asia (which uses gravity to deploy irrigation water through tiers of terraces).

The sawah system compensates for the general paucity of nutrients in tropical soils by storing nutrients in water, rather than in the biomass, as occurs with swidden. Consequently, the sawah system is relatively insensitive to the inherent quality of the soils. It brings nutrients to the irrigated terraces in the water, and the formation of blue-green algae fixes nitrogen. The gradual movement of the water aerates the soil and improves its tilth. The control of water application allows more careful timing of farm operations, compared with rain-fed farming, a factor which helps to boost yields to levels more than twice those of rain-fed systems, at around 2.5 tonnes of cereals/hectare. Although yields may fall in the first 2 years of the operation of such a system, they become remarkably stable thereafter. This reflects the fact that the inherent fertility of the soil is less critical for farming success than is control of the water supply.

In contrast to the low carrying capacity of the swidden system (Table 3.1), the sawah system is associated with the densely populated lower Nile valley and low-income Asia. It has exhibited a great capacity to absorb increases in population, a feature described by the anthropologist Clifford Geertz (1963) as 'agricultural involution'. Basically, this describes the capacity of the system to go on yielding higher levels of output which amply compensate for additional inputs of labour. For example, instead of sowing seeds by broadcasting, they can be grown on the banks of the fields and then carefully transplanted onto the terraces. Although this is much more labour-intensive than simple broadcasting of the seed, it yields sufficient extra output to compensate for the additional work effort. Or again, human 'scarecrows' can so reduce crop losses to birds and pests that the extra labour effort expended is adequately compensated. A final example of the labour absorptive capacity of the sawah system is the application of night-soil to raise yields through the organic enrichment of the land. Meanwhile, where the growing season permits, multiple cropping can take place, with up to three harvests reaped during a single year.

Scientific farming systems

The fourth stage of the intensification cycle, widely known as the green revolution, squeezes an even greater output from the irrigated system, through the application of quick-return scientific inputs. It involves a package of improvements which, in addition to irrigation, includes high-yield varieties of seeds, fertilizer, pesticides and crop storage. There is also a need for adequate all-weather roads to ensure the timely arrival of inputs and the dispatch of the harvested crops, but further discussion of this aspect is delayed to a fuller analysis of the relationship between transport and

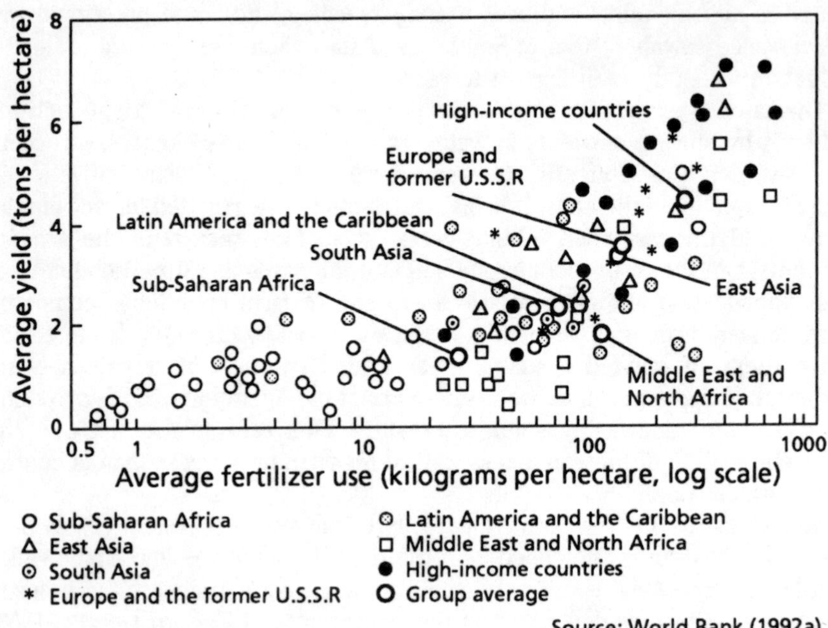

Fig. 3.2 *Fertilizer inputs and cereal yields 1989, by country*

development in the large low-income Asian countries, in Chapter 4. The green revolution system is capable of yielding around 4 tonnes of cereals per hectare and it is estimated to have been adopted by almost half of Asian farmers by the 1980s.

The fifth and final stage in the agricultural transition requires the infrastructure of a sophisticated distribution system and is associated with the developed countries. This system is also typically characterized by a rapid increase in the size of farms and a high level of field mechanization, reflecting the increasing scarcity of farm labour as the economy matures. It may typically employ barely 2 per cent of the workforce (compared with more than 80 per cent in farming under traditional society). Cereal yields reach 7 tonnes per hectare under this system, but it demands modern delivery systems, including efficient wholesale and retail outlets.

But the delivery system pushes up the fraction of the labour force engaged in the food sector to one-fifth of the workforce, and also further depresses the energy-efficiency of the system. According to Pimental *et al.* (1973), a system in stage five of the agricultural transformation uses 6 times as much energy per calorie of food produced as under the swidden system. There are also, of course, problems with soil erosion and environmental pollution (Makhijani 1975), which are discussed in the context of low-income Asia, a region whose less abundant land availability has pushed it further along with scientific agriculture than Africa.

Even aside from qualms about increased dependence on toxic chemical inputs and finite sources of energy (Fig. 3.2), green revolution techniques are unlikely to be warranted where, as in much of sub-Saharan Africa south of the Sahel, population growth has far from exhausted the prospects for expansion through additional land (Table 3.4). A study by Binswanger and Pingali (1988) criticizes many African farm schemes for intensifying production, on the grounds that extensive rain-fed farming systems are superior where land/labour ratios are still favourable.

Agricultural intensity and land scarcity

The agro-climatic population density index

The intensification of yields traced in the third and fourth stages of the sequence outlined above is appropriate for densely settled areas like low-income Asia and the lower Nile valley, but not for the more sparsely settled areas of sub-Saharan Africa. Failure to recognize that the conventional green revolution approach is premature in such areas is one reason behind the disappointing record of many World Bank and UN agricultural schemes in sub-Saharan Africa.

In order to illustrate this point, Binswanger and Pingali (1988) calculated the physical potential for food production under different levels of technology and produced an index of resource adequacy by relating the food production potential to population. Their index is called the agro-climatic population density. It is calculated by dividing each country into agro-ecological cells and estimating the maximum number of calories of food production that could be sustained under three technologies (low input or shifting cultivation, intermediate input with perennial rain-fed agriculture, and high input with green revolution techniques).

The index is forward-looking, so that the projected population of each country by 2025 is then divided by the estimates of potential calorie production under intermediate technology (the level projected for much of sub-Saharan Africa in the first quarter of the twenty-first century) to produce a standard index of population density. Expressed verbally, the index of agro-climatic population density measures the number of people per million calories of food potential. A low density on this index is 100 or fewer people per million calories of food potential, whereas a high density is 250 people per million calories or more.

The index yields some interesting results (Table 3.4), with the desert country of Egypt ranking as high-density, along with countries which contain large semi-arid areas such as Niger and Ethiopia, bordering the Sahel, and

Table 3.4 *Population density by climate, selected countries, 1990–2025*

Climatic category	Low density[a]	Medium density[b]	High density[c]
Humid lowlands country	Guinea Bissau (2028) Malaysia (2091) Liberia (2051) Equatorial Guinea (2088) Zaire (2080) Congo (2109)	São Tomé and Principe (2041) Sierra Leone (2054)	Bangladesh Mauritius
Share of sub-Saharan countries in total sub-Saharan population	8.7%	0.9%	0.3%
Mixed climates[d] Country	Ivory coast (2038) Chad (2041) Bahamas (2086) Madagascar (2041) Argentina (2123) Cameroon (2045) Brazil (2119) Zambia (2066) Angola (2071) Central African Republic(2114) Gabon (2147)	Gambia (2031) Zimbabwe (2032) Togo (2033) Ghana (2036) Tanzania (2033) Benin (2040) Costa Rica (2097) Guinea (2072) Sudan (2065) Mozambique (2062)	Kenya Rwanda Barbados Burundi Comoros Mauritania Ethiopia India Nepal Nigeria Uganda Malawi
Share of sub-Saharan countries in total sub-Saharan population	12.8%	22.1%	46.3%

Arid or semi-arid climates country	None	Mali (2027)	Niger Somalia Lesotho Afghanistan Pakistan Egypt Namibia (1988) Senegal (2006) Mexico (2019) Botswana (2023) Burkina Faso (2024) Swaziland (2024)
Share of sub-Saharan countries in total sub-Saharan population	0%	1.8%	7.1%

Source: Binswanger and Pingali (1988)

Note: In each climatic category, countries are ranked by projected population density in 2000. Data are unavailable for the sub-Saharan countries of Cape Verde, Djibouti and Seychelles.

[a] Having less than 100 people per million kilocalories of potential production by 2025. Figures in parentheses denote the year a country is expected to reach a density of 100.

[b] Having 100 people per million kilocalories of potential production currently or by 2025. Figures in parentheses denote the year a country is expected to reach a density of 250.

[c] Having 250 people per million kilocalories of potential production currently or by 2025. Figures in parentheses denote the year a country that has not yet reached a density of 250 is expected to do so.

[d] Includes climates with mostly intermediate rainfall and countries with both high- and low-rainfall zones.

Kenya. It is clearly no coincidence that the examples of environmental erosion in Chapter 2 are drawn from these parts of Africa. Despite their modest total populations (8 million and 25 million, respectively) and large geographical size (1.26 million and 0.58 million sq. km, respectively), Niger and Kenya are more densely settled than Bangladesh, which has 111 million people in 144 000 square kilometres (World Bank 1993b). This is because the great extent of the semi-arid areas in the African countries reduces their population carrying capacity. Nevertheless, the index of agro-climatic potential suggests that land is likely to remain plentiful elsewhere in Africa away from semi-arid and arid areas such as the Sahel and lower Nile Valley into the second quarter of the next century.

The swidden farming system is well adapted to the condition of low agro-climatic population density because it maximizes the use of plentiful land and minimizes the use of scarce labour. An ample supply of land in relation to labour renders the expensive inputs of fertilizer associated with the green revolution unnecessary. In contrast, densely settled regions of South Asia, like the Punjab, which straddles the border of India and Pakistan, have rural land costs of up to $33 000/hectare, but labour is abundant. Such regions do need to economize on the land input through the use of green revolution techniques, which take advantage of surplus rural labour to intensify output per hectare.

Assuming a sustained population growth in Africa of around 3 per cent annually, Binswanger identifies three conditions (of high, medium and low agro-climatic population density) by 2025. The high density of more than 250 persons per million kilocalories of potential production is just below the level achieved by India and Egypt in the 1980s. The low-density index of less than 100 people per million kilocalories of potential production compares with an index of 127 people per million kilocalories for Thailand in 1980. The specific countries which fall into each category are shown in Table 3.4, which also distinguishes between the countries according to climate (humid lowland, semi-arid and mixed).

The regions which are projected to reach a high density by 2025 include Mauritius, southern Nigeria, the East African Rift (Kenya, Uganda, Burundi, Rwanda, Ethiopia, Malawi) and the Sahel. They will by then account for about one-half the population of sub-Saharan Africa and will need to be applying land-intensive green revolution techniques. But around one-fifth of the region's population will still live in countries whose agro-climatic population density will be low, so that swidden cultivation will remain an appropriate farming system. These countries are mainly in the humid lowlands of Central Africa and the West African rainforest, excluding southern Nigeria. Finally, about one-third of the countries of sub-Saharan Africa (with one-quarter of the projected population) will fall into the intermediate class by 2025. They are mainly in eastern and western Africa and farmers there will be making the transition to more intensive farming systems. The process of land-use intensification within an area will begin with the most fertile and accessible sub-regions.

Policy implications

Binswanger argues that many African agricultural projects, often sponsored by the UN and the World Bank, forced African farmers into intensifying land use prematurely. He concludes that the reluctance of African farmers to respond to the new opportunities which such schemes offered them was due neither to ignorance nor to stubbornness. When infrastructure is provided and soils are good, African farmers quickly switch from subsistence to cash crops, and use new varieties with resistance to drought, pests and diseases. But they are, correctly, not interested in intensive farming methods as long as land is plentiful and cheap.

But there are two constraints in addition to land availability on the intensification of farming in much of sub-Saharan Africa. One of these is a lack of skilled management and cheap labour with which to construct the infrastructure for irrigation. The capital investment required by such schemes in sub-Saharan Africa tends to be several times higher than in Asia. The other constraint results from the urban bias of government policies which, as noted in Chapter 2, have squeezed returns to peasant farmers and thereby depressed incentives to abandon subsistence agriculture in favour of commercial farming.

Transportation plays a catalytic role in the overall progression to a more intensive land use. This is because, as illustrated earlier with reference to the Machakos district of Kenya, it allows farmers to move out of subsistence by opening up markets so that they can become more specialized. The productive potential of the more fertile areas results in their being opened up first; the resulting increase in profitability then attracts in-migration, which eventually drives up land values. This triggers the shift towards a more intensive cultivation which, in a continental perspective, can be expected to diffuse out from sub-Saharan Africa's most favoured farming regions, like the volcanic uplands of the Rift Valley, into the mixed climatic zones of eastern and central Africa and finally into the humid lowlands such as the Zaire Basin.

A green revolution for sub-Saharan Africa

Harrison (1987) has developed a rural-driven development strategy which recognizes the failures of national policy and institutions in Africa. He seeks to provide solutions which, in the first instance, can be implemented by peasant farmers with minimal dependence on state assistance and imported inputs. The two subsequent steps assume some improvement in the basic constraints, such as easier access to imported inputs. But the initial step entails the application of techniques by local communities which provide high returns at low cost and with little risk.

Step one: high return techniques with low cost and low risk

The principal improvements which fall within this category pay particular attention to soil conservation, and entail terracing and windbreaks. Soil-loss estimates in south-eastern Nigeria suggest that a 3 centimetre reduction in topsoil can cut yields by 40–50 per cent. Consequently, methods which reduce soil loss are likely to yield a substantial improvement in agricultural performance. One practical measure to arrest soil loss entails the digging of a narrow trench along the contour line, and piling the soil on the uphill side to retard soil movement.

A similar goal can be achieved by laying trash from the harvest along the contour line and allowing soil terraces to form behind the resulting ridge. An additional terracing measure involves laying stones along the contour line, this time in order to arrest moisture flushing across thin soils. The barriers along the contours also collect rain-borne leaves and dead grass, thereby enriching soil nutrient content. In order that the barriers can be aligned with the slope contours, hose-pipes have been deployed to act as improvised spirit levels. Although the stone barriers take up 1–2 per cent of the land area, they have been shown to raise output by up to 50 per cent in Bukina Faso, where they were originally developed.

In the drier semi-arid areas, wind erosion poses a greater threat to soil loss than rainfall and may be countered by the use of windbreaks. For example, in the Sahelian region of Niger, the rate of soil loss per hectare was reported to be 20 tonnes in 60 kilometres per hour winds. By planting trees at 1-metre intervals in lines some 100 metres apart, crop yields can be boosted by 20 per cent, even allowing for the loss of farmland to the trees. Moreover, the firewood derived by culling the trees provides an added bonus, where hydrocarbon fuels are prohibitively expensive and local bushes and trees have been removed (as reported in Chapter 2, on Ethiopia). An alternative source of fuel wood (and animal fodder) can be secured by planting trees around houses. This is also likely to bring savings in labour, because it is estimated that women spend increasing amounts of their time – up to half of each day – scavenging for fuel wood to cook food.

A second means of reducing wind erosion is the practice of alley cropping, where leguminous trees are planted with crops in alleys almost 5 metres wide. The trees are pruned during the cropping season and allowed to shoot up 4 metres in height during the dry season. The trees are then pollarded back to 1–2 metres as the new crop season begins. This system yields fuel wood and also stakes and twigs for mulch/fodder, while protecting soil. Under these conditions, recordings in south-western Nigeria suggest that maize yields can be more than doubled, reaching 2 tonnes per hectare – a very high level for a rain-fed farming system.

Step two: low-cost techniques with some import dependence

The second set of improvements places greater demands on national institutions. It involves the development of new crop varieties and cheap labour-saving devices. Little conventional green revolution research into high-yield varieties is suited to African conditions, given the widespread absence of irrigation, the nature of the soils and poor economic management. But some progress has been made – for example, at the Ibadan Research Centre in Nigeria, which developed a new hardy domestic cassava. The new cassava variety matures in 8 months rather than 12, and requires one-third of the weeding of traditional plants. This is an important consideration, since weeding is thought to comprise one-third of the total work input on such farms. The new variety of cassava is also more resistant both to disease and to drought. Finally, yields from the new crop are 2 to 3 times those of traditional varieties of cassava.

Such new local crop varieties can be quickly distributed, because they are cheap to obtain, robust and low-risk in contrast to the new varieties grown in the more well-tended and therefore 'artificial' breeding stations, whose products often fail under actual African farm conditions. A second relatively inexpensive means of improving farm output is the purchase of cheap wells, pumps and grinders. By giving greater control over the timing of farm tasks such as planting, weeding and harvesting, they contribute to improved agricultural performance and also save labour.

Such improvements may be especially important for women farmers, who comprise up to three-quarters of all farmers in some African countries where the men migrate to cities in search of work. Data for Zimbabwe suggest, for example, that women farmers work more hours per day on the land than men (8 compared with 7 hours) and then spend an additional 5 hours daily on household chores, compared with barely 1 hour for men. In such a context the need for women to spend many hours gathering fuel wood or carrying water (before embarking on cooking inside a smoky kitchen) is clearly an urgent problem which the social forestry schemes noted under stage one are addressing.

Step three: initial stages of Asian-style green revolution

In the more densely settled areas of those countries which can achieve reasonable macroeconomic management and sustain effective local institutions, a case can be made for the introduction of high-yield variety crops with irrigation, fertilizers and the whole package of green revolution techniques. Zimbabwe adopted such a programme in hitherto neglected rural areas where black farmers produced maize.

The initial cost of the Zimbabwean scheme was relatively high, at \$150/hectare, and farmers needed adequate access to credit in order to make the initial outlay. Nevertheless, yields tripled from 0.6 tonnes per hectare and the country's total production of maize jumped to 1.8 million tonnes in 1985, yielding a surplus of 1 million tonnes for export. But an important prerequisite for this success was the establishment of prices at levels which gave the farmers an incentive by yielding an adequate return.

Conclusion

Many African countries do not yet need the more sophisticated land-intensive farming systems that more densely settled parts of the developing world require. The post-independence problems of African countries have underlined the inappropriateness of large-scale, technologically complex farm projects. Such projects are required at fairly late stages in the intensification of land use, when the opportunities for smaller improvements have been exhausted and the large-scale coordination of a sophisticated infrastructure, such as that associated with the application of green revolution techniques to newly irrigated areas of Asia, is the only option. Even then, such a system presupposes a reliable supply of both inputs (some imported) and skills which some African governments have found difficult to provide.

Instead, the population–land ratio outside the semi-arid regions calls for measures to improve farm output which are low-cost, low-risk and locally self-sufficient. Such measures are aimed at conserving land and improving yields. They can also buy time in which to revitalize the government, establish sound macroeconomic policies and improve the efficiency of institutions. Thereafter, the provision of external communications to the more productive regions will trigger a sequence of interactions between in-migration, technology and land use, capable of raising incomes with minimal environmental damage. But such developments presuppose the successful political resolution of the tribal and ethnic frictions which have beset most African countries since independence (Sandbrook 1986).

But there is another lesson to be learned by Africa from the past 3 decades, which concerns the rate of population growth. As the example of Egypt clearly shows (Chapter 2), large land-intensive projects such as those associated with the Aswan High Dam may give a false sense of security which masks the underlying problems created by rapid increases in national population. Moreover, a bias towards large-scale schemes may be also ill-judged, because these incur high maintenance and rectification charges, offering second-best solutions compared with more modest schemes which seek to improve the use of existing farmland resources.

Despite some undoubted economic benefits arising from the increased size of domestic markets and higher rural population densities, rapid population growth is likely to depress average living standards and retard the rise of incomes and welfare, compared with a steady fall in fertility. It has been estimated that if the sub-Saharan African region halved its birth rate between 1990 and 2020 to 21 births/1000, then it would also halve the volume of extra food required. It is much more likely, however, given the formidable political obstacles to improved government, that the deceleration in population growth will be much slower, and consequently that the eradication of poverty will be much more protracted.

PART 3

Income inequality: low-income Asia

4

Income inequality: distance and development in low-income Asia

The focus on rural development continues within this chapter, because a majority of the population in the large low-income Asian countries (China, India, Pakistan and Bangladesh) lives in rural areas (Table 4.1). (It should be noted, however, that China experienced a remarkable surge of urbanization in the 1980s, discussed in Chapter 5.) Like sub-Saharan Africa, the large Asian countries also initially neglected agriculture, but efforts to remedy that unsatisfactory state of affairs started earlier and met with more success. The origins of rural neglect in both low-income regions can be traced to the adoption of a policy of forced industrialization.

It may be recalled from Chapter 1 that Myint considered the resource endowment of the large Asian countries (large domestic markets and a shortage of arable land) to favour development based upon labour-intensive industrialization, along the lines of the East Asian development model. But strategic considerations led India and China especially among the large low-income Asian countries to opt for an autarkic (i.e. self-sufficient) industrialization strategy. This requires the rapid expansion of heavy and chemical industry (HCI), because at low levels of per capita income that is the sector of the economy which is most deficient (compare columns 1 and 2 of Table 4.2). Yet the skill and capital which HCI-led growth requires are exactly those factors of production which are scarce in low-income countries. Put more formally, HCI-led growth is not in line with the comparative advantage of low income, which lies in labour-intensive light manufacturing such as textiles.

The problems of HCI-led growth are examined in more detail in Chapter 5. This chapter concentrates on the consequences of that strategy for income distribution and the rural sector. As in sub-Saharan Africa, the costs of a mistaken development strategy have fallen to a disproportionate degree on the rural poor. A strategy of autarkic industrialization requires a country to reinvest each year a very high fraction of GDP, and this leaves less available for consumption – a sacrifice which most of the population can ill-afford. Nevertheless, both the Indian and Chinese governments did seek to equalize incomes, for ideological reasons, even though that aim clashed with the goal

Table 4.1 *Economic and social indicators for selected low-income Asian countries*

Country	Area (million sq. km)	Population (million)	Cropland/ hd (ha)	Population growth (%/yr)	Population in cities 1991 (%)	GDP growth 1980–91 (%/yr)	Per capita income ($1991)	Literacy (%)	Life expectancy (yrs)
Bangladesh	0.144	110.6	0.08	2.2	18	4.3	220	35	51
China	9.561	1149.5	0.08	1.5	22	9.4	370	73	69
India	3.288	866.5	0.20	2.1	25	5.4	330	48	60
Pakistan	0.796	115.8	0.18	2.8	29	6.1	400	35	59
Sri Lanka	0.066	17.2	0.11	1.1	21	4.0	500	88	71
Sub-Saharan Africa	23.066	488.9	0.33	3.1	22	2.1	350	50	51

Sources: World Bank (1993b), except cropland/hd = WRI (1992)

Table 4.2 *Changing structure of production with per capita Income (%GDP)*

Per capita income ($ 1980)	180	300	500	1000	2000	4000	7300
Agriculture	48.0	39.4	31.7	22.8	15.4	9.7	7.0
Mining	1.0	5.0	6.6	7.7	7.5	6.1	1.0
Manufacturing	10.0	12.1	14.8	18.1	21.0	23.6	28.0
Light industry	6.0	7.6	8.3	9.2	9.8	10.1	11.9
Heavy and chemical industry	4.0	4.4	6.5	8.9	11.2	13.5	16.1
Chemicals	1.5	2.4	2.9	3.5	3.8	4.1	3.5
Non-metallic minerals	0.4	0.5	0.8	1.1	1.1	1.3	1.5
Basic metals	0.5	0.5	0.9	1.3	1.6	1.8	2.1
Metal products and machinery	1.6	1.0	1.9	3.1	4.7	6.4	9.0
Construction	4.0	4.4	4.9	5.5	6.1	6.7	7.0
Utilities	6.0	6.7	7.4	8.1	8.8	9.3	10.0
Services	31.0	32.4	34.6	37.8	41.2	44.7	47.0

Source: Syrqin and Chenery (1989, pp. 20 and 32)

of industrial self-sufficiency. Lucas (1988) describes the economic system adopted by China as command socialism, distinct from the command capitalism of India's mixed economy.

The present chapter begins with a systematic examination of the role of agriculture in economic development, noting how that role has been revised from the mid-1960s as evidence mounted concerning the adverse social and economic consequences of rural neglect. An important barrier to rural development is the inadequacy of transport and communications in many rural areas. This occupies the second section of the chapter, which explores the effect of distance on development. It is argued that the emerging transport systems of developing countries have often served to bridge space, rather than to open it up: the transport system links major cities but leaves most rural areas as islands of neglect. The third section of the chapter examines the prospects in low-income Asia for a rural-led development strategy based on the green revolution. The actual achievements of rural areas are then evaluated in the light of the emerging environmental and egalitarian critiques of the green revolution.

Agriculture, economic development and poverty

The changing role of agriculture in economic development

In the 1950s, many development economists were pessimistic about the prospects for agriculture, especially in the land-scarce countries, and this led them to advocate rapid industrialization. It was thought that land shortages implied the early onset of diminishing returns on investment in farming, whereas industry held almost unlimited opportunity for raising productivity through the constant application of new technology. Farmers were also seen as resistant to change: small farmers seemed to be irrational, because they failed to respond to price incentives, while the larger farmers with ample incomes tended to be complacent. They supported the status quo and opposed rural reforms.

Nor was land reform seen to offer much hope for any improvement in densely settled Asia, in contrast to land-abundant South America (Grindle 1986). For example, the 1960 Indian census revealed scope to redistribute 65 million hectares, if the maximum holding were restricted to 10 hectares (Mellor 1976). Although that would have been enough land to raise the size of all small plots to 2 hectares, and also to provide landless labourers with 2-hectare farms, the continued growth in population would quickly eliminate any gains. Such a pattern of land redistribution would leave all farms too small to be viable. Rapid industrialization therefore appeared necessary in order to draw surplus labour from the land, so that farms could become larger and more productive. A rise in farm size allows rural incomes to match the rise in urban incomes.

The key obstacle to industrial development in the 1950s was thought to be a shortage of capital. It was assumed that heavy industry would raise the economy-wide rate of reinvestment and thereby lead to faster economic growth. This is because, unlike farming or light industry, heavy industry is capital-intensive and has a high propensity to save (Lewis 1954). The high returns from investment in HCI were expected eventually to spill over into more consumption-oriented activity (Balasubramanyam 1984). Such views prompted both the Indian and Chinese governments to espouse autarkic industrialization strategies in the 1950s, and to assign agriculture a secondary and supportive role. For example, the Indian government reduced its budget spending on agriculture from 15 per cent to only 13 per cent between 1950 and 1960, despite the fact that two-thirds of the country's labour force and half its GDP originated from that sector.

China and India both accelerated their HCI 'drives' into HCI 'big pushes' in the 1950s. A 'big push' attempts to simultaneously establish both the intermediate industries, like steel, along with their downstream markets, like shipbuilding and engineering. This draws on the ideas of Rosenstein-Rodan (1943) and is designed to maximize the economies of scale. Demand for each product can be larger if the upstream and downstream plants are built simultaneously: they thereby achieve lower production costs and become more competitive. In this context, farming was regarded as a convenient vehicle to absorb surplus labour, while also supplying a stream of capital, labour, cheap food and materials for the emerging manufacturing sector.

Johnston and Mellor (1961) identify five key functions for agriculture during economic development. First, agricultural exports supply the foreign exchange with which to import the machinery initially required to establish new industries. Second, it provides the government with the tax revenues needed to build up the national infrastructure. A third role is to supply physical inputs to factories, many of which draw on agricultural supplies in the early stages. The fourth function is to provide a growing surplus of cheap food for urban consumption, while the fifth function is to release labour in order to meet the demand for workers created by the expanding industrial sector. Given this set of functions, it is not surprising that governments in Asia, as in sub-Saharan Africa, should be tempted to squeeze more and more from agriculture. Indeed, the efforts of the Indian government to extract the rural surplus led Lipton (1977) to formulate his urban bias thesis.

But the HCI big pushes were impractical and overstrained domestic implementation capacity. There were insufficient skilled workers, engineers and managers to efficiently undertake such large-scale industrial expansion. In the case of China, the shortages of heavy industrial products led to the ill-fated Great Leap Forward in 1958. Shortfalls in the HCI sector were to be remedied from small backyard production facilities established on the communes. This policy assumed there was surplus labour in the rural sector which could be diverted to produce the required shortfall from the large factories without loss of farm output. This assumption proved mistaken, and an estimated 15 million died through starvation, as a result of labour being withdrawn from food production (Riskin 1988).

The Indian planners were more cautious than their Chinese counterparts, but they only succeeded in postponing the basic problem of inadequate skills until the mid-1960s, when the costs of rural neglect were clearly demonstrated. Depressed production and income in the agricultural sector led to a decline in profits and in household saving, which reduced the rate of investment. A fall also occurred in government revenues, along with a sharp drop in demand for industrial products. Manufacturing experienced rising costs, declining returns and falling investment. Economic growth was disappointingly slow. This undesirable outcome underscored the vital role which agriculture plays as a *complement* to industry.

The failures of the Indian and Chinese big pushes demonstrated that farmers are consumers as well as producers. Farmers provide an important market for the manufacturing sector so that slow income growth in the rural sector implies lost demand for manufactured goods. Such findings coincided with research which indicated that peasant farmers are responsive to market incentives, provided the new methods carry moderate risks – a not unreasonable requirement, by individuals close to the economic margin (Schultz 1964). In addition, the diffusion to Asia of green revolution techniques pioneered during 1945–65 in the Philippines and Mexico added to the improved outlook for rural investment.

Other shortcomings of the HCI-led strategy also became evident. Not only did it require a very high rate of investment, which meant lower levels of consumption, but also the return on investment fell as it was applied to progressively more capital-intensive uses. Capital efficiency declined and labour-intensive light manufacturing was starved of funds – its share of GDP actually shrank in the case of India. Between 1950 and 1970, Indian textiles production deteriorated from being a competitive exporter, to an ailing inefficient industry. Moreover, the slow growth of labour-intensive manufacturing combined with the low rate of job creation in capital-intensive HCI to cause the rate of unemployment to grow faster than expected. The fraction of the rural population living below the poverty line in India is estimated to have increased through the 1960s, from 38 per cent to 56 per cent (Mahendra Dev *et al.* 1992).

Poverty and employment

Measurement of absolute poverty requires estimates of the per capita expenditure required for a minimum standard of nutrition and other basic necessities, along with the cost of participating in day-to-day society. Such figures must then be adjusted for differences in purchasing power between countries, i.e. for the fact that a dollar spent in one country may buy more of a given basket of goods than in another country (services, for example, are often

Table 4.3 *Regional incidence of poverty in developing countries, mid-1980s*

Region	Total poor (million)	% population poor	Infant mortality (deaths/ 1000)	Life expectancy (yrs)	Primary school enrolment (%)
South Asia	520	52	172	56	74
China	210	20	58	69	93
Sub-Saharan Africa	180	47	196	50	56
Latin America	70	19	75	66	92

Source: World Bank (1990a)

much cheaper in low-income countries than in mid-income ones). Table 4.3 uses World Bank (1990a) estimates of poverty in the mid-1980s. It sets the poverty line at $370/year, with extreme poverty defined as $275/year or less.

The World Bank identified some 1.116 billion people worldwide living in poverty – one-third of the total population of the developing countries. Of these, almost half were estimated to live in extreme poverty, a situation which could be relieved by redistributing just 3 per cent of the annual expenditure by developing countries on consumption from the two-thirds living above the poverty line to those living below it. But even if this redistribution were politically feasible and actually achieved, the problem would not disappear, because poverty is 'relative' to some norm in society, which rises as welfare improves. The World Bank (1990a) calculated that whereas $350 defined poverty in countries where average per capita income was $600, the level rose to $650 in mid-income countries with a per capita income of $2400, and to $1600 where the average income was $4800. Consequently, as per capita incomes rise, absolute poverty may shrink but relative poverty may decline more slowly, or even increase.

Kuznets (1971) argued that the relationship between per capita income growth and income inequality traces an inverted U-shape, so that income inequality at first rises as per capita income increases, and then falls at the higher levels of income. Although World Bank data for five levels of income in the mid-1970s did reveal such a pattern, there were anomalies. For example, Latin American countries tended to have relatively high levels of income inequality, as did the African countries for which data were available (Table 4.4) whereas the level of inequality in China was relatively low. Moreover, the East Asian dragons have since confirmed that rapid economic growth, rising per capita incomes and an equitable distribution of income are not incompatible.

Nevertheless, urban incomes are typically more than twice those of rural areas, even in China, and income inequality may be expressed in strong

Table 4.4 *Income inequality variations*

	Income distribution Quintiles		
	Lowest	Highest	Ratio
Latin America			
Argentina	4.4	50.3	11.43
Brazil	2.0	66.6	33.30
Chile	4.5	51.3	11.40
Colombia	2.8	59.4	21.21
Costa Rica	3.3	54.8	16.61
Ecuador	1.8	72.0	40.00
Mexico	4.2	63.2	15.05
Panama	2.0	61.8	30.99
Peru	1.9	61.0	32.11
Trin and Tob	4.2	50.0	11.90
Uruguay	4.4	47.5	10.80
Venezuela	3.0	54.0	18.00
Average	3.2	57.7	21.1
East Asia			
China	7.0	39.0	5.57
Hong Kong	6.0	49.0	8.17
Indonesia	6.6	49.4	7.48
Korea	6.5	45.2	6.95
Malaysia	3.5	56.0	16.00
Philippines	3.9	53.0	13.59
Singapore	6.5	49.2	7.57
Taiwan	8.8	37.2	4.23
Thailand	5.6	49.8	8.89
Average	6.0	47.5	8.7
Other			
Egypt	4.6	48.4	10.52
Hungary	10.0	34.0	3.40
India	4.7	53.1	11.30
Israel	8.0	39.0	4.88
Ivory Coast	2.4	61.4	25.58
Kenya	2.6	60.4	23.23
Mauritius	4.0	60.5	15.13
Morocco	4.0	49.0	12.25
Portugal	5.2	49.1	9.44
Spain	6.0	45.5	7.58
Sri Lanka	6.9	44.9	6.51
Tunisia	6.0	42.0	7.0
Turkey	2.9	60.6	20.90
Yugoslavia	6.6	41.4	6.27
Average	5.3	49.2	11.7
Overall average	4.8	51.7	14.1

Source: Sachs (1989)

regional differences. For example, Bangkok accounts for one-fifth of the Thai population and has an average per capita income of \$2500, compared with \$350 for the two-fifths of the population resident in the rural north east. Such figures are consistent with the important finding that participation in the modern sector of the economy is a prerequisite for poverty alleviation. This is why the exclusion of so many people from labour-intensive employment by HCI-led growth is so unfortunate. A second key condition for poverty alleviation is the targeting of social funds at the very poor in order to reduce hardship, with special attention being given to assistance with health care and education.

Prior to the 1970s, many low-income countries attempted to improve health with prescription drugs and centralized hospitals, which proved to be an expensive and urban-biased form of treatment. A growing number of countries have since shifted towards primary health care targeted at children and child-bearing women, with measures such as:

- the provision of clean water (since 80 per cent of infections are spread by water);
- vaccination, for as little as \$5/child, against childhood diseases (which account for one-third of infant mortality);
- contraceptives provision (since one-tenth of women die during a pregnancy and an added bonus is that a reduction in family size helps to ease poverty);
- child care (basic education concerning nutrition and hygiene, such as the need to sterilize equipment used to cut the umbilical cord, or to treat diarrhoea with fluids and salt rather than by withholding water);
- the provision of local health workers at village level;
- the purchase of only those drugs which are cheap to secure and easy to administer.

In education, there is a need to focus scarce public resources on primary education. In India between 1960 and 1980, for example, primary education expanded from 61 per cent to 79 per cent of the population, compared with rises in secondary education from 20 per cent to 30 per cent and in tertiary education from 3 per cent to 8 per cent. Wastage rates in primary education are lower than in higher education, yet even so, half those entering fail to complete 4 years. The costs per student of primary education are one-twelfth those of secondary education. Higher education costs 90 times as much to provide per student as primary education, and many university students go abroad and fail to return, while others merely use their credentials to secure a job for which they are over-qualified.

There are also social benefits from primary education which stem largely from the opportunity which the basic literacy and numeracy skills provide for improving the quality of life. The teaching of basic health care (nutrition and hygiene) has a very substantial pay-off in terms of lower population growth and improved family health. Primary education for girls is especially useful in improving their status and family care skills. Finally, most tasks are unlikely to require greater knowledge than that secured in primary school, especially in the low-income countries, while the low cost of primary education, compared with secondary and tertiary education, renders it easy to make such provision widely available, including in rural areas. But this assumes an adequate communication system has been established which, as the next section shows, has rarely been the case for rural areas.

Transportation and economic development

Taaffe *et al.* (1963) have developed a useful six-stage model which traces the emergence of a transportation system in a developing country. The model represents an ideal pattern, so that different parts of the transportation system may be at different stages in the sequence. Basically, what the model shows is how sharp differences in accessibility emerge which favour a few well-connected urban centres, while leaving much of the countryside isolated.

Fig. 4.1 summarizes the sequence, beginning with the transport system for a traditional society which would comprise scattered small ports dotted along the coast, each serving a hinterland of a few kilometres radius. Land transport comprises meandering bush trails linking largely subsistent villages, and movement is far easier by boat than over land. In stage two, penetration lines emerge as a small number of ports are selected to open up the hinterland to establish law and order and to facilitate trade. The penetration lines may give access to interior forts and administrative centres; or they may take the form of railway lines where large volumes of bulky materials, such as minerals, are to be traded; or they may make possible the export of crops from plantations or peasant plots.

The second stage is the critical one in the evolution of the transport system because it establishes centres of high accessibility (cities) unrelated to the local natural resource base through the operation of agglomeration economies. The agglomeration economies represent the gains from the clustering of economic activity, which include lower infrastructure costs and access to much of the skilled labour which the country possesses. Once a centre acquires a head start, it grows faster than rival centres because of the advantages of agglomeration economies.

The consolidation of the initial advantage is shown in stage three of the Taaffe model (Fig. 4.1) as feeder networks sprout from ports and inland centres, further augmenting *their* growth, while most of the small ports go into decline. In stages four and five lateral interconnections emerge, as feeder lines extend towards each other and link up. This further enhances the accessibility of the larger centres and reinforces their attraction. The final stage traces the appearance of a national trunk transport network which connects only the very largest centres and consolidates still more their prowess.

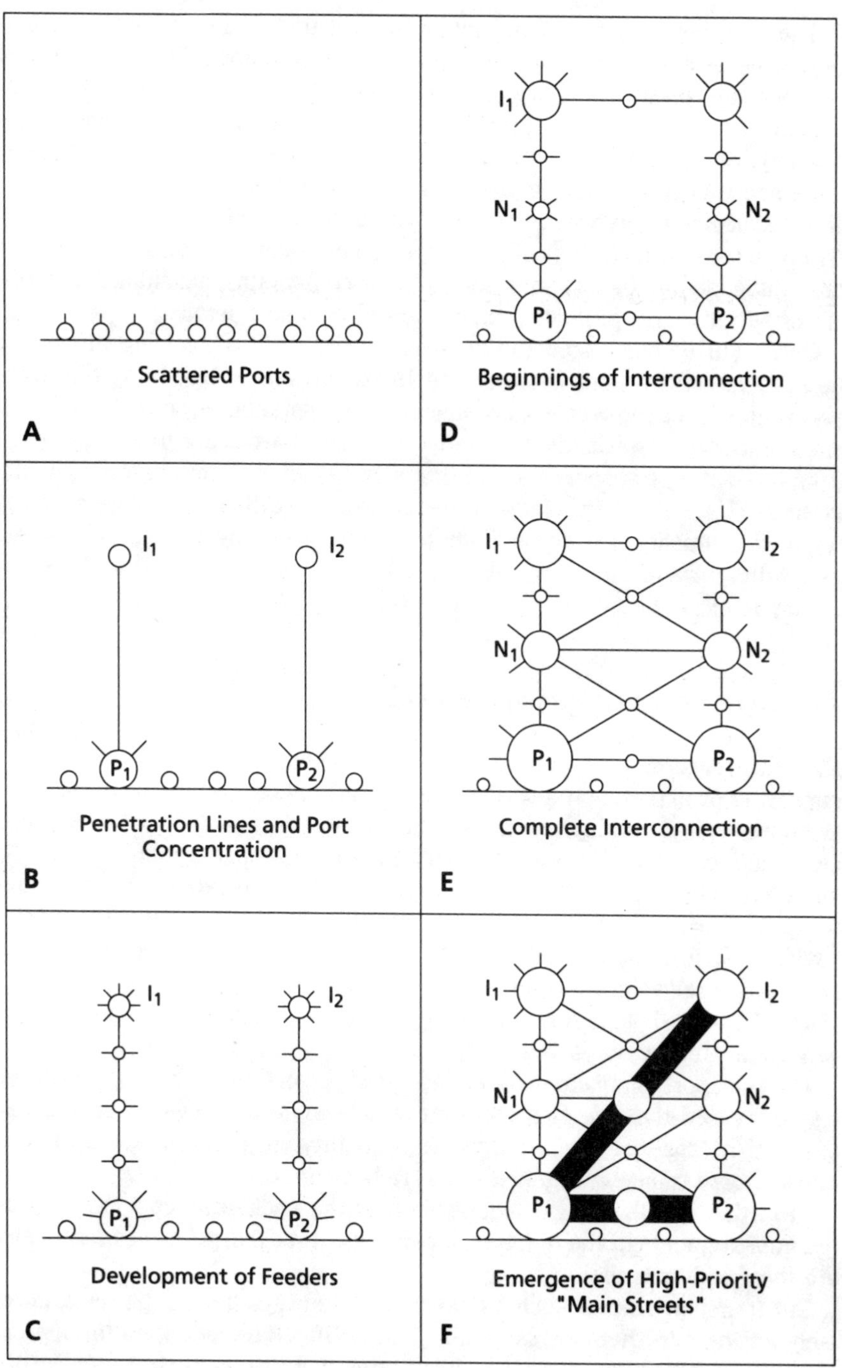

Source: Taaffe *et al.* (1963)

Fig. 4.1 *Transport network evolution in developing countries*

The evolving transportation system sets in motion two important spatial processes – concentration and regional specialization. The first process concentrates tertiary and secondary activity in a few very large centres, leading to an urban system described as primacy. In a primate spatial system, one city, or a handful of very large cities, accounts for a disproportionately large share of economic activity. This issue is analysed further in Chapter 6 with particular reference to hyper-urbanization in Latin America, the most highly urbanized major developing region. The second process leads to areal differentiation, as regions specialize in their comparative advantage as suppliers of specific farm products, manufactured goods or services.

Gould (1970) has traced the process in terms of the evolving modernization process of Tanzania between the 1920s and 1960s. He shows that transport routes, in effect, *bridge* space to link the larger centres, so that rural areas are subdivided into islands of poverty. Most rural areas are many kilometres from all-weather transport and are therefore decades, if not centuries, behind in time. The evolution of the transport network therefore implies rising regional inequality, a process likely to be reinforced by development strategies with a bias towards urban-based industry which neglect agriculture – as is very much the case with HCI-led development.

Transport modes and movement costs

Historical records demonstrate the dramatic fall in costs associated with improving transportation and also the cost advantages which well-connected locations enjoy. Taking as a starting point pre-industrial transportation, eighteenth-century American colonists found it cheaper to ship goods 5000 kilometres by sea than 15 kilometres over land (Savage 1966). This is because land transport was then by packhorse, which was slow and carried a very low payload. Hence, coal mined in south Lancashire at that time cost three old pence/hundredweight to mine, but a haul of just 15 kilometres doubled the price of the coal. It is not surprising therefore that pre-industrial transport was dominated by ships and ports.

The transport innovations of the industrial revolution cut haulage costs and also increased the size of the objects which could be moved. For example, the British canals of the late eighteenth century cut costs to one-third those of packhorse transport, while the first railway trains cut costs *vis-à-vis* canals by a further two-thirds, i.e. to one-ninth of the packhorse cost. Subsequent rail improvements in the nineteenth century cut rail freight costs to one-thirtieth the packhorse cost.

But transport modes such as canals and railways benefit places immediately adjacent to them (Kolars and Malin 1970). Railways open up opportunities in a narrow swathe a few kilometres wide along their route: further away from the line of rail, unless all-weather roads are provided, the high costs of manhandling goods to the track quickly eliminate incentives to trade.

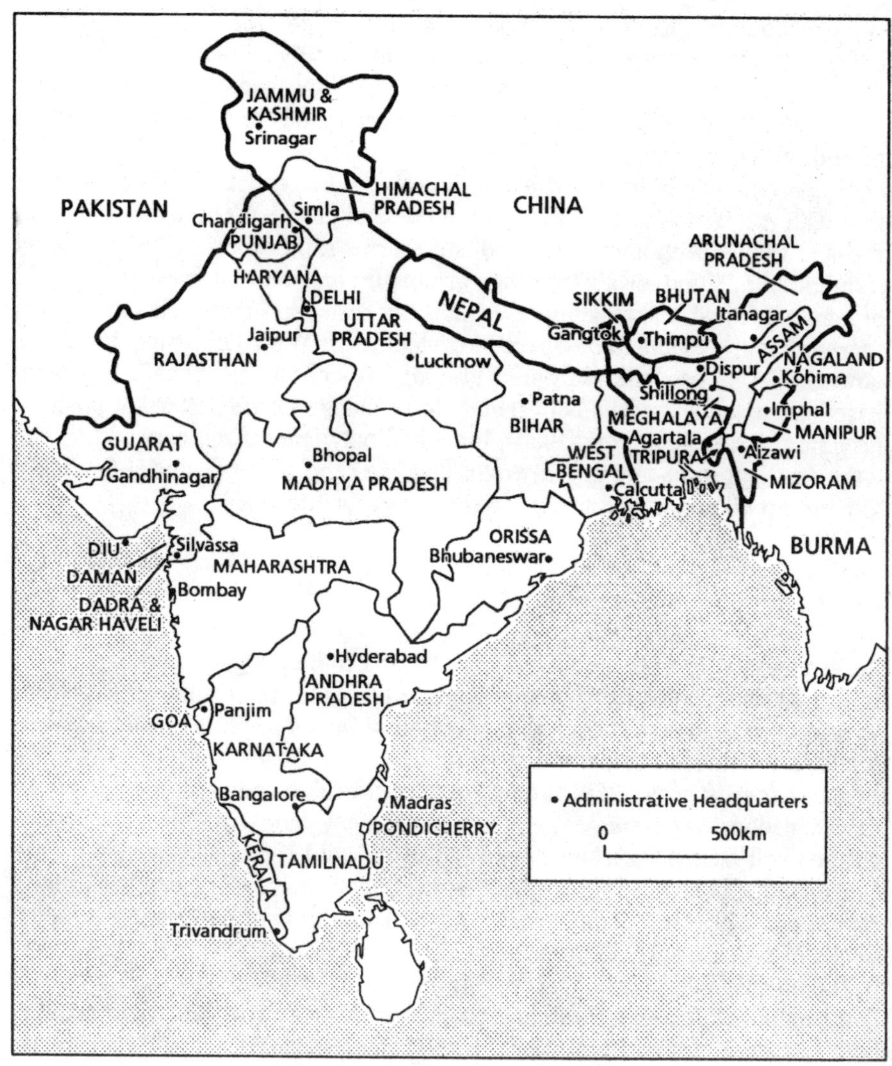

Fig. 4.2 *Principal cities and states of India*

It is for this reason that railways and motorways bridge space, while all-weather rural roads open it up. The neglect of all-weather roads in favour of railways and highways will polarize economic opportunities.

India during the period of its HCI big push (1956–65) provides an example of the modal bias of transport (Owen 1968). Preference was given to commerce and industry (of which Bombay alone produced one-third of the country's output). Consequently, although 80 per cent of the Indian population lived in the countryside, rural areas received only 35 per cent of the total investment between 1950 and 1964. In the case of transport investment, some 75 per cent went to intercity modes of transport, mainly to the inefficient state-owned railways.

This urban bias in transport expenditure was driven not only by the industrial policy: it was also amplified by the political process. The planners found it easier to deal with a few large industrial firms rather than millions of small farmers. In addition, urban pressure groups (unions and businessmen) could organize political lobbying more easily than dispersed rural folk: they were therefore more visible to city-based legislatures. In the countryside, the more powerful large landlords accepted low food prices, to the benefit of urban-based union leaders and businessmen, in exchange for the latter not pushing for land reform. By the mid-1960s, India had only one million cars and trucks – the mode of transport which would have had the greatest spatial flexibility to open up *all* areas, not just to boost access for the cities.

Rural transport

Rural transport in India and China has tended to be irregular and unreliable. In the 1960s, one-third of Indian villages were more than 8 kilometres from an improved road, and the inhabitants of more than four-fifths of villages were required to traverse paths which were unfit for vehicles in order to reach a surfaced road (Owen 1968). Many roads and bridges became impassable in the monsoon. Yet improvement has been slow, and two-thirds of Indian villages still lacked an all-weather road in the mid-1980s (Owen 1987). One consequence of such conditions is that profit margins are drastically slimmed by the costs of transporting goods by animal or human porterage over dirt-tracks to reach a surfaced road.

The unreliability of road transport is one factor which makes farmers reluctant to abandon their subsistence production in favour of cash-crop production. For example, farmers urged to produce vegetables and milk are likely to find their produce perishes before reaching the market. Even basic services such as vets, doctors and teachers become expensive to supply to rural areas, because of the length of time such individuals must spend completing their journey by walking to the village.

Willbanks (1972) shows that if distance from an all-weather road is plotted against modernization indices like land productivity, it produces an L-shaped curve. This means that there is a rapid fall-off in the application of new farming techniques with distance from a surfaced road: remoter villages are unlikely to exhibit much if any farming change. Without surfaced roads to move inputs in, and outputs out on time, the package of activities which comprise the green revolution is useless. Local storage and crop processing are also required to free farmers from the extortion of local hauliers and/or merchants. Finally, extension services are a vital part of the green revolution package, but Owen (1967) estimated that one officer typically got through only 300 man hours of instruction per week, compared with 600 with better roads, and much more with adequate radio communication.

The neglect of rural areas (and also of labour-intensive manufacturing in the cities) was associated with growing income inequality in the large low-income Asian countries. By the mid-1970s the ratio of the income of the richest quintile to the poorest quintile in India was 11 (Table 4.4), almost 3 times that of Taiwan. Although the Chinese figure is a more respectable 7 (and this appears to be borne out by social data such as life expectancy), urban incomes were outstripping rural incomes with the urban–rural income ratio widening to 3 to 1. Moreover, anecdotal evidence (Chang 1991) suggests that the Great Leap Forward and Cultural Revolution greatly increased hardship in rural areas. In this context, by the 1970s, the relatively slow economic growth, low investment efficiency and income inequality associated with HCI-led growth combined with the promise of green revolution technology to encourage investigation of rural-led development.

Prospects for rural-led development

Mellor's thesis

In a book entitled *The New Economics of Growth*, published in 1976, Mellor made the case for a rural-led strategy of economic growth in India based on an intensification of land use. He estimated that if the annual growth rate in agriculture could be increased to 4 per cent from the less than 3 per cent which had been achieved, the sector would give a very strong stimulus to the economy because of its still relatively large size and very beneficial multiplier (or linkages). He calculated that full employment could be reached within little more than a decade with such an agricultural growth rate, compared with three times as long at prevailing rates of growth. Moreover, if the rate of population growth were to decelerate to 2 per cent, this would halve the time required to achieve the full employment target. As noted earlier, full employment is one of the two conditions for easing poverty.

Mellor argued that the diffusion of green revolution technology could be substantially extended, because less than half of India's irrigation potential had been tapped and the irrigated area could easily be expanded using labour-intensive means. By the mid-1970s, green revolution techniques were being applied to almost 100 per cent of the millet-growing areas and 85 per cent of the wheat-growing areas, but to barely 30 per cent of the rice-growing areas (Chakravarti 1973). The difficulties with rice meant that a rapid eastward expansion of wheat occurred in the Ganges valley, displacing rice.

Mellor estimated that farming would directly supply two-fifths of the increased jobs created by faster agricultural growth, with the remaining jobs coming indirectly through the multiplier. He focused on the different multipliers associated with the three principal groups of rural worker, each of which accounted for about one-third of the rural population. The three groups comprised landless labourers, small landholders (around one-third of whom had so little land as to be little better off than the landless) and the larger farmers (classified as those with more than 2 hectares of land). The incidence of poverty is twice as high among landless labourers as among other rural groups, according to Mahendra Dev *et al.* (1992): they experience the highest unemployment and have the lowest literacy (25 per cent).

Mellor argues that the agricultural multiplier is maximized if the largest farmers take full advantage of the green revolution techniques. Paradoxically, the multiplier will be minimal if the rural poor increase their income, because they will consume most of any extra food output themselves. Mellor estimated that in the 1970s the poorest rural quintile would spend 60 per cent of their extra income on food grains, compared with an increased expenditure of only one-sixth for the second wealthiest rural quintile. The second richest quintile would also seek to improve its diet as its income increased (almost one-third of any additional income would go on non-food grain products, mainly milk and vegetables).

The trickle-down effects of spending by the wealthy farmers to the poorer rural groups lies at the heart of Mellor's thesis. He estimated that, of the extra revenues received by the larger farmers, about one-quarter would go on the purchase of green revolution inputs and one-tenth on additional farm labour, leaving almost two-thirds as reimbursement to land and capital. The benefits of that higher income would trickle down in three important ways. First, extra work would be created for the landless labourers, as the wives and children of the richer farmers withdrew from the workforce. Second, the improvements in diet of the wealthier farmers would create a demand for milk and vegetables, products which can be intensively produced and which are well-suited to the smaller plots of the poorer farmers. Finally, the increased expenditure of the wealthier farmers on non-food products would be on labour-intensive goods which could be made in local workshops by those landless labourers unable to find work on the larger farms.

The smaller farmers would tend to diversify into the production of milk and vegetables, which require more labour/hectare than food grains and also yield a higher income per hectare. This would reduce their need to seek work on large farms and thereby open more farm work to landless labourers. Traditionally, the latter had been unemployed for at least one-quarter of the year, usually between crops. More labour would be required through the rest of the year at a higher wage to apply the additional inputs which scientific farming requires. Indeed, the larger farmers might mechanize some tasks to cope with labour shortages in the peak harvest period. Finally, those landless labourers who failed to find farm employment would have widening work opportunities in a range of labour-intensive rural industries. Firms would be set up to supply basic equipment for the green revolution, such as tube wells and simple pumps. These firms could later diversify into other engineered products (household fittings, for example) as demand for pumps and wells matured.

Infrastructure would also improve because the incomes of the wealthier farmers would be taxed to improve rural water supplies, schools and communications. The construction of such facilities would also boost rural employment for the landless labourers. These gains would also help reduce the cost of delivering food to the cities and of carrying manufactured goods to the rural areas by between one-fifth and one-third. But care would need to be taken to ensure that the savings arising from increased rural wealth were not siphoned away to the cities as the national banking system expanded into the more prosperous rural areas.

According to Mellor (1976), rural-led development would be less capital- and energy-intensive than HCI-led growth, and it would also produce less atmospheric pollution. For example, the incremental capital output ratio (or ICOR, which compares the extra unit of investment required to produce an extra unit of output) for green revolution agriculture has been estimated at around 1, compared with around 6 for Indian HCI-led growth in the 1960s (Mundle 1990). Finally, rural-led development also has positive welfare implications. It can improve rural productivity, incomes and diet, thereby helping to slow or even reverse the growing income gap between the cities and the countryside.

Consequences of the green revolution

Although India failed to achieve the gains which Mellor hoped for (Mundle 1990), China made spectacular progress when the Dengist reforms were first applied to the rural sector in 1978–84. Whereas farm productivity had declined under the commune system 1957–75 (Rawski 1979) and output was only sustained by a sharp increase in the per capita hours worked, the Dengist reforms heralded a spectacular 8 per cent growth in annual farm output. Basically, the granting of increased security of land tenure to farm families at last resulted in full advantage being taken of the rural infrastructure improvements made under the commune system.

There is also evidence from China of substantial multiplier effects, as families diversified into services and small-scale manufacturing. Rural incomes rose sharply, while city dwellers experienced some decline in their real incomes because of rising food prices and job competition from a rapid surge of rural in-migration. Zweig (1987) estimates that the urban–rural income gap in China narrowed from 3 to 1 to 2 to 1 through the early-1980s.

But world-wide, the green revolution techniques have not fully lived up to the highest expectations set for them. Moreover, they have been criticized for widening income differentials, both within the countryside (between larger farmers and landless labourers) and between regions (areas with favourable ecology and good road networks have left less well-endowed regions behind). There have also been environmental criticisms, which centre around the application of chemical inputs and the misuse of irrigation water.

Social criticisms

The diffusion of green revolution techniques in South Asia did not lead to rapid acceleration in food production, but it may well have averted a slow-down which might have occurred in the absence of the new technology (Griffin 1988). One reason for this is that green revolution techniques have not diffused uniformly through rural areas: they tend to favour the largest farmers and the more accessible regions with suitable growing conditions.

Within the more accessible and productive areas, green revolution techniques have conferred their benefits on the largest farmers, although they are supposedly scale-insensitive (i.e. the techniques can be deployed effectively irrespective of the size of farm). The net effect of these trends has been to amplify income inequality ***within*** rural areas, as well as between regions whose natural resource endowment differed in terms of adaptability to the new technology.

Quizon and Binswanger (1986) found that the larger Indian farmers were able to innovate first, and that they benefited from the fact that grain prices remained high as the government took advantage of the extra output to curb food imports. By the mid-1970s, however, the increase in grain output was reflected in lower prices, which adversely affected those smaller farmers who had been unable to adopt green revolution techniques earlier because of a lack of credit. Those small farmers who belatedly adopted the new techniques captured little of the higher income which the larger innovators had enjoyed. Yet, whereas the larger farmers had scope to adjust their labour and other inputs to offset any fall in crop prices, smaller farmers lacked such flexibility.

At a national scale, green revolution techniques have not been equally applicable throughout India, and their diffusion has therefore widened the income gap between rural regions. By the mid-1980s they had been applied on about one-third of the cultivated area (Owen 1987). The states of Bihar, Madhya Pradesh, Uttar Pradesh and Rajasthan, which account for two-fifths of India's population, have lagged (Figs. 4.2 and 4.3). While ecological drawbacks and the absence of credit provide one set of reasons, another is the lack of adequate transport. As noted earlier, in the mid-1980s some two-thirds of Indian villages still lacked an all-weather road (Owen 1987). Nor can land redistribution ease the lot of the rural poor, because only 5 million of India's 80 million farm households own more than 6 hectares of land. The redistribution of common property lands provides no solution either, because continuous encroachment has led to their annexation by private individuals and/or in their environmental deterioration (Mahendra Dev *et al.* 1992).

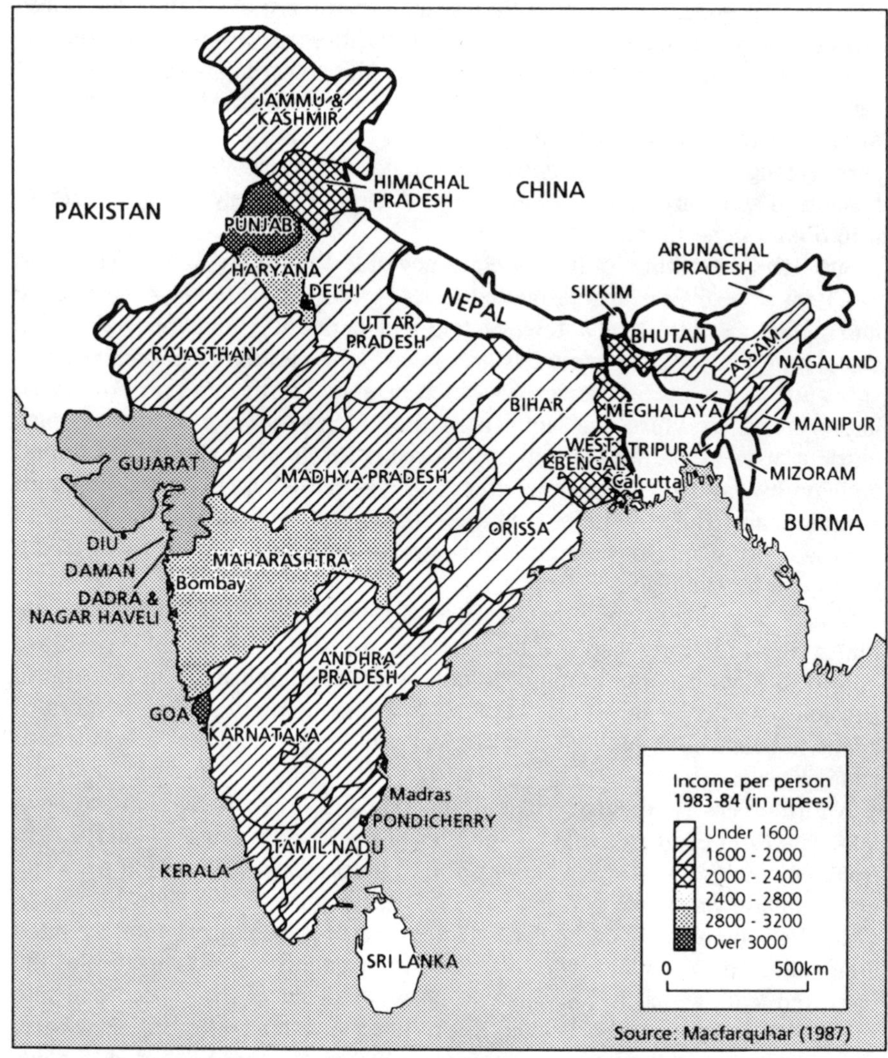

Fig. 4.3 *Variations in income distribution, India mid-1980s*

Environmental critiques

Concern has been expressed about the use of chemical inputs such as pesticides and nitrogenous fertilizers in developing counties (Repetto 1989). The problem is particularly acute where government subsidies for their use are high. This is because farmers receiving subsidized inputs have less incentive to economize on their use, or to seek biological substitutes (such as natural predators). Hence, the policy of the Indian government of setting food prices low and then subsidizing farm inputs like fertilizer in order to compensate, distorts the application of fertilizer and pesticides towards imprudent use and can lead to poor management. Such subsidies were initially deployed in order to encourage small farmers to adopt green revolution techniques, but they persist long after such incentives are needed. Moreover, the economic benefits of subsidized inputs often accrue disproportionately to larger farmers who do not need them.

Subsidies contribute to the low efficiency of fertilizer use. Some 50 per cent of chemical fertilizers are recovered in the crops, with much of the residue finding its way into the water systems and polluting them. Subsidies also discourage the use of organic fertilizers which, in addition to providing nutrients such as manganese and zinc, which chemical fertilizers do not do, also improve soil structure. In sandy soils they increase water retention, curb nutrient loss, and buffer the soil against rising levels of alkalinity, acidity and toxicity. In clay soils, organic matter improves porosity, raises soil moisture retention and thereby helps reduce erosion. In fact, yields are likely to improve with a mix of chemical and organic fertilizers – or at least, falling yields can be stabilized.

Similar arguments apply to pesticide subsidies which, in addition to the human health hazard, also risk becoming ineffective: more than 400 pests are known to have become immune to one or more chemicals. The removal of subsidies would be beneficial because it would help encourage farmers to adopt a balance of biological, chemical and crop rotational measures to combat pests.

A further environmental problem affecting the green revolution is salination. The FAO estimates that one-quarter of the world's irrigated land suffers from salinity (*Economist* 1992), with the highest ratios in India (36 per cent, or 20 million hectares) and Pakistan (27 per cent, or 3.2 million hectares). China has about 7 million hectares (7 per cent of the country's total irrigated area) affected. Normally, where sufficient irrigation water is available, the dissolved salts which it contains can be washed out of the topsoil and into the rivers, or they can be flushed down into the water table. But elsewhere, the saline water cumulates in the water table, as on the Asian flood plains. Over time, the saline water in the water table has risen close enough to the surface in some areas to be drawn by evaporation into the crop roots, with disastrous consequences. Correction requires the construction of drainage channels to take the saline water away. This is expensive, however, with one Pakistani estimate showing that it costs five times as much to construct the required drainage channels as it does to provide irrigation water in the first place (compare the experience of Egypt, in Chapter 2).

The expansion of irrigation to accommodate the green revolution has also raised controversy, as the example of the two large dams (Sardar Sarovar and Narmada Sagar) proposed for the Narmada River in north-west India shows. The benefits from the Sardar Sarovar scheme include an additional 1.9 million

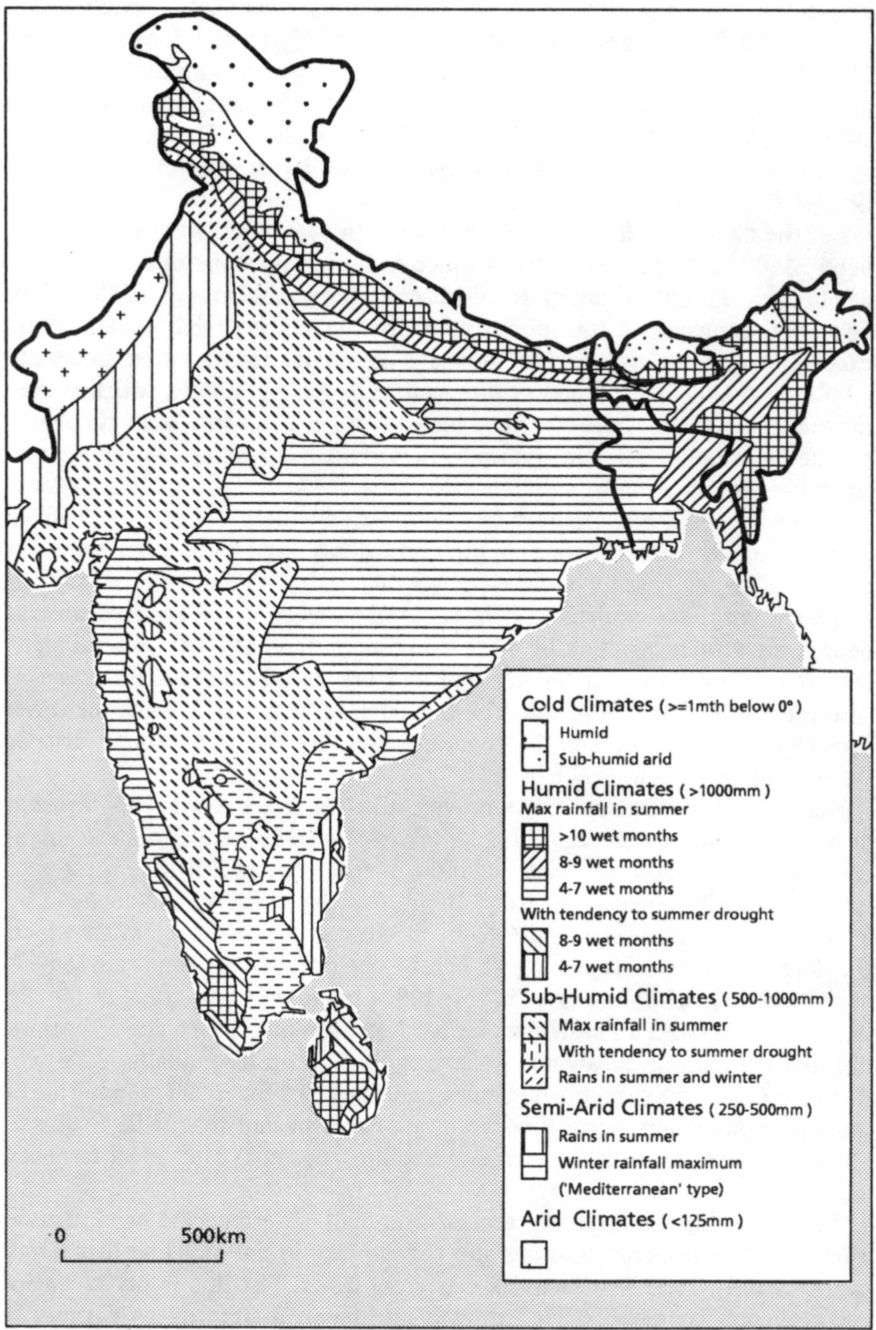

Fig. 4.4 *Climates of South Asia*

irrigated hectares which would accommodate 340 000 farm families and create 700 000 extra labouring jobs (Mikesell 1992). In addition, some 1450 MW of hydroelectric power would be provided, as well as improved rural and urban water supplies. The proposed Narmada Sagar dam would add about 200 000 irrigated hectares and almost 1000 MW of power. The projects would mainly benefit some 10 million people in the dry north-western state of Gujarat (Fig. 4.4).

But the dams will displace 150 000 people (mostly poor tribal groups) and flood 94 000 hectares (two-fifths of which is cultivated, while most of the rest has been degraded). Although the displaced persons and loss of habitat might be amply compensated, the opponents of the scheme claim that the total costs exceed the benefits, especially if the dams are poorly implemented, like some others have been. Resolution of the issue calls for techniques which permit rational comparison of the costs and benefits of environmental change, a topic pursued in more detail with reference to sustainable development of mineral economies in Chapter 9.

Meanwhile, as the battle over the merits of extending irrigation intensifies in the irrigated Asian floodplains, the rain-fed farming zones suffer from soil erosion. One estimate for India suggests that 12 billion tonnes of topsoil are lost each year, the equivalent of 2 centimetres per hectare and capable of depressing crop production by 40 million tonnes annually. The rain-fed areas are cropped by June and the cracked and dry soil is vulnerable to erosion with the onset of the monsoon. The monsoon rains are so intense that even protective mud banks along the contour are of little use in slowing rainwater run-off and curbing topsoil loss.

Hedges of hardy vetiver grass provide a better check against soil erosion in India than low mud contour banks (*Economist* 1986) while reafforestation can also help. Barely one-sixth of India is forested, and at recent rates of loss (1.5 million hectares annually) the remaining forest will be depleted by 2010. The main pressure comes from an internal demand for timber which the forest service is replacing at barely one-fifth of the required rate. A social forestry scheme begun in the 1980s made little progress on public lands, but it thrived on individual farms. It is estimated that a single hectare of rapidly maturing eucalyptus generates $500 of profit annually, with two-thirds of the income from wood sales and one-third from fuel wood. This is more profitable than most other crops and has appealed both to large absentee landowners (due to the minimal tending required) and small farmers, alike.

The persisting stress on the rural environment underlines the need to draw more people into manufacturing and service jobs in the cities so that farms can increase in size and incomes can grow faster. But this is not so easily accomplished because HCI-led growth is difficult to reform, as Chapter 5 demonstrates. Moreover, such a wholesale redistribution of population merely substitutes one set of environmental problems for another, as Chapter 6 shows.

Conclusion

The role of agriculture underwent a re-evaluation in low-income Asia during the 1960s, and spectacular rural progress was made in China during 1978–84 when the Dengist reforms were applied. Progress in South Asia has been less spectacular, with the notable exception of the Punjab and Haryana wheat-growing region (Fig. 4.2). Whereas in China the rural–urban income gap closed sharply in the early 1980s, Indian rural incomes continued to fall behind urban ones into the 1980s, when they were about one-quarter the urban average (Mundle 1990). Poverty in India remained strongly concentrated in rural areas (Bhargava 1987), and in the mid-1980s around half the Indian population remained poor, with one-third classed as 'extremely poor' (Crook 1991).

HCI-led growth has been an important cause of the slow reduction of poverty in low-income Asia. By using capital inefficiently and attentuating the growth of labour-intensive employment, it leads to a very slow absorption of labour into the commercial economy. Yet the provision of paid employment is one of two prerequisites for rapid poverty alleviation. The second is the targeting of state assistance at the poorest in society through the subsidized supply of primary education and basic health care.

The opportunity provided by the green revolution to compensate for the misconcieved HCI-led strategy has fallen well short of Mellor's hopes for India, but by the 1990s government policies were no longer sustainable in either the rural or the industrial sectors. An important reason for this is that public sector finances could no longer accommodate the policy of subsidies, much of it related to agriculture. In rural areas, the depletion of natural resources in the face of mounting population pressure suggests that a more rapid rate of urban growth, such as that experienced by China in the 1980s, is required to pull people off the land into labour-intensive urban employment. But this requires a radical shift away from HCI-led growth, which neither China nor, especially, India, have found easy to achieve. What the new policy might be and what obstacles its implementation faces are discussed in Chapter 5.

5

The costs of autarkic industrialization in India and China

Industrial policy options

The large low-income Asian countries are densely settled and the resulting scarcity of land means that industry needs to absorb increasing numbers of workers from the countryside. Those who remain behind as farmers then have the opportunity to increase their welfare by intensifying production on ever-larger farms. But the postwar development strategy which the large low-income Asian countries favoured is not an efficient way of achieving this objective. This is because that strategy attempted to achieve a high level of industrial autarky (self-sufficiency), which requires heavy investment in capital- and skill-intensive HCI. This results in the neglect of labour-intensive agriculture and light industry in which the comparative advantage of the large low-income countries lies. Moreover, the labour-intensive nature of the latter sectors offers the fastest prospect of absorbing all workers into the modern economy, a key prerequisite for poverty alleviation.

This chapter examines the industrial strategies of India and China in detail. It begins by setting the policy choice in the context of the industrial strategies open to the developing countries. In particular, the inward-oriented (autarkic) import substitution industry (ISI) strategy is contrasted with the outward-oriented export-led growth of the East Asian development model (first introduced in Chapter 1). Within the inward-oriented strategy, a useful distinction may be made between the very high level of autarky sought by the large (market-rich) developing countries, and the more constrained ISI options of the smaller (market-poor) countries.

The second and third sections of the chapter then analyse the highly autarkic industrial policy (which can also be described as HCI-led growth) chosen by India and China, respectively. In each case, the disadvantages for the economic growth rate, income distribution and investment efficiency are outlined. The differing obstacles to industrial policy reform which each country encountered are then discussed and the emerging regional problems of the more dynamic Chinese economy are analysed.

The case for an active industrial policy

During both the 1930s and the 1939–45 World War, when many Asian and African countries were still colonies, the more advanced self-governing Latin American countries were forced to industrialize via ISI. This is because the Depression of the 1930s had cut their export earnings so that they were unable to buy manufactured goods from the industrial countries. This process of import substitution continued through the 1939–45 World War because the industrial countries oriented their manufacturing capacity towards the production of armaments.

Trading conditions eased in the late 1940s and engendered more optimism concerning the economic prospects for the developing countries. For example, the process of industrialization was expected to continue in Latin America, reducing the region's external dependence. In addition, industrialization would strengthen the new bourgeois and industrial classes, at the expense of the landed oligarchy. The Latin American oligarchy historically derived its power from the control of export agriculture and the dominant economic role which that sector played. The oligarchy provides an example of internal colonialism. Class power realignments were expected to bring democracy, which would help to redistribute wealth and speed the elimination of poverty. Finally, income inequality would shrink at the global scale, as industrialization closed the technical gap with the developed countries.

The initial postwar expectations proved too optimistic and this led Prebisch (1963) and others to reassess Latin America's options. Prebisch believed that international trade favoured the accumulation of capital by the industrial countries (see Chapter 8 for a fuller discussion). He concluded that the governments of the developing countries needed to pursue voluntary ISI and to build up their manufacturing behind protective trade barriers. The trade barriers would be reduced once modern competitive industries had been established and all countries could engage in international trade on equal terms.

A second set of arguments in favour of government intervention to promote industrialization relates to the acquisition of technological knowledge. As 'late industrializers', the developing countries do not need to develop their own technology, but they do need to close the technology gap with leading countries. According to institutional economists like Amsden (1989) this is not easily achieved by market forces: rather, it requires state intervention in order to protect 'infant industries' and also to provide incentives and subsidies to encourage domestic firms to invest.

Domestic firms (the 'new entrants') need assistance in bearing the risks of closing the technology gap with experienced, established firms in the industrial countries. This is because the head-start of the established producers enables them to use predatory pricing (the supply of goods at low prices in order to drive would-be new competitors out of business), and this poses a formidable entry risk. The new trade theory of the 1980s provides further reasons for state intervention to promote industry, but the reasons are more appropriately discussed with reference to Korea and Taiwan in Chapter 11.

Inward-oriented industrial strategies

Basically, ISI uses import statistics to identify those products whose demand is nearing the scale required to support efficient domestic production. In the early stages, the most suitable goods are those like textiles and footwear, which make little use of scarce skills and capital. Another early group of industries comprises bulky products using local materials, such as cement and pottery.

As home demand rises towards a size at which domestic production becomes viable, a licence is granted to a single local producer in order to prevent the scale benefits from being dissipated among too many local entrants. When the new plant is ready to operate, imports are restricted either by raising the import tariff to make the import more expensive (and less competitive), or by an outright ban on the hitherto imported product. Domestic producers may also be given incentives such as low interest loans and tax holidays, in order to spur investment. Over time, as the 'infant' producer becomes more efficient, both the barriers against imports and the incentives to new entrants are removed.

In practice, ISI creates numerous difficulties. It causes some initial deterioration in the balance of payments, as capital goods are imported to construct the new factories. Even after the start-up of local production, import content may be high – this is because the manufacturing process often involves the assembly of a wide range of components, many of which are unavailable in a developing economy. Customs revenue is also lost when imports are curbed, so that ISI involves a transfer of resources from the government to the favoured new firms. Taxpayers may also lose out in their role as consumers, because all too often the new product initially costs significantly more than the imported product did, and is of a lower quality.

When ISI is widely applied, as was the case in many developing countries, the economy may become severely distorted. This is because ISI discriminates in favour of production for (protected) domestic markets against production for export markets, which is far riskier. In many cases the effective rate of protection (which measures the level of protection enjoyed by that part of the manufacturing process which is undertaken domestically (i.e. *excluding* purchased inputs)) may exceed 100 per cent. Further distortion arises when infant industry incentives such as tax holidays and low-interest loans encourage capital-intensive production at the expense of job creation.

Meanwhile, as indicated in Chapters 2 to 4, an active industrial policy discriminates against agriculture and those (usually labour-intensive) manufacturing subsectors which are either not protected from import competition, or receive lower levels of protection.

At the root of the problem with ISI lies the fact that it creates rents (returns in excess of those required by a competitive producer) because of the restriction on foreign and domestic competition. In fact, it may become more remunerative for industrialists and their workers to expend energy competing for and extending the life of those rents, than to seek ways of improving their competitiveness. There is also scope for corruption if the rents are shared with the politicians who create them and the civil servants who administer them. All too frequently, industrial policies have been captured by these vested interest groups which benefit most from them. Such groups then pose a formidable obstacle to reform, no matter what the political system.

The rent-seeking interest groups tend to favour over-valued exchange rates because their protection insulates them from any loss of competitiveness such exchange rates imply, while the prices of imported components (and also consumer goods) which they may wish to use become cheaper in local money terms. Meanwhile, as the exchange rate strengthens, unprotected activities such as agriculture, mining and exporting in general become less competitive. Yet, the protection and subsidies are not withdrawn from the infant industry as it matures – rather, they tend to continue. Consequently, the maturation of 'infant industries' has often taken many decades, rather than the five to eight years required if the (discounted) long-term benefits from the new industries are to compensate for the medium-term costs of infant industry support (Bell *et al.* 1984).

The protracted maturation of infant industries places a substantial economic burden on the primary sector of the economy (agriculture and mining). That burden proves increasingly difficult to bear as per capita income rises and the size of the primary sector shrinks *vis-à-vis* that of the expanding protected (and inefficient) manufacturing sector. The appreciation of the exchange rate also weakens the primary sector. The resulting squeeze on profitability causes fiscal and trade deficits, which engender stop-go macroeconomic responses: economic growth becomes both more erratic and slower (Table 5.1).

As evidence emerged through the 1970s of the long-term cumulative damage which poorly applied ISI policies can inflict, the strategy came under fierce attack from orthodox economists. For example, Greenaway and Chong (1988) argue that policies which insulate economies from external competition lower economic performance compared with outward-oriented policies. In a blistering attack on development economics, Lal (1983) identified ***dirigisme*** (high levels of state intervention, as advocated by Prebisch and other structuralists) as a key obstacle to rapid economic growth. In his view, if sophisticated OECD governments are hard-pressed to intervene effectively

Table 5.1 *Features of competitive and autarkic industrial policy*

	Competitive	Autarkic
Industrial policy		
Market impact	Market-conforming	Market usurping
Incentives duration	Tapered off	Renewed
Economy openness	High	Low
Sectoral targeting	Sequenced HCI	Most HCI
Macro policy		
Overall stance	Pragmatic orthodoxy	Structuralist
Budget balance	Tight	Deficit accumulation
Exchange rate	Competitive	Over-valued
Some policy outcomes		
HCI maturation rate	5–10 yrs	Decades
Non-HCI viability	Strong	Weak
Foreign exchange	Unconstrained	Constrained
Turning point	Rapid arrival	Delayed arrival
Income distribution	Equitable	Skewed
GDP growth rate	High	Modest
GDP sustainability	High	Erratic
Inflation	Moderate	High

Auty (1994)

in their own economies, it is absurd to expect the equivalent of eighteenth-century governments in the developing countries to play an even more ambitious role.

Meanwhile, the ISI strategy has been attacked from another direction. Economists like Krugman (1979) and Beenstock (1983) have challenged the Prebisch assumption about the strength of the industrial countries. They speculate that the industrial countries may have increasing difficulty in holding their own against the newly industrializing countries of East Asia. But to ascribe, as for example Balassa (1985) does, the success of the latter economies entirely to outward-oriented policies (implying there is no scope for interventionist industrial policies) is too simple.

Outward-oriented industrial policy

The merits of an active industrial policy in developing countries remain controversial. This chapter certainly provides clear evidence that an autarkic (i.e. strongly inward-oriented) industrial policy adversely affects long-term economic growth. An outward-oriented or competitive industrial policy may well be highly beneficial (Amsden 1989, Wade 1990), although conclusive evidence is still awaited (Ranis and Mahmood 1992). Unlike an inward-oriented industrial policy which mutes efficiency incentives, a competitive industrial policy is market-conforming because it repeals competition only temporarily. It is sometimes called a dualistic policy (Ohno and Imaoka 1987), because it simultaneously strives to maintain the competitiveness of established sectors, while encouraging new industrial sectors with emerging comparative advantage.

A competitive industrial policy provides a package of market information, assistance with technology acquisition, subsidized credit, tax breaks and trade incentives for new entrants to set up infant industries. But in return it demands that the favoured firms must mature rapidly. Maturity requires firms to compete internationally, with no subsidies other than those needed to compensate for the use of high-cost inputs from the next wave of protected industrial sectors. They must also be able to make in-house technical adjustments that enhance the productivity of the newly installed plant.

A competitive industrial policy is associated with the East Asian model of economic development outlined in Chapter 1. To recap, that model identifies four stages in the economic transition from a low-income to a high-income country (Kuznets 1988). Each stage lasted about 1 decade in Korea and Taiwan. An initial protective stage of industrialization (often referred to as primary import substitution) is supported by transfers of foreign exchange and subsidies from the primary sector. It is followed in the second stage by a phase of export-led growth, encouraged by fiscal and financial assistance to exporters. The exports tend to be light manufactured goods such as textiles, in line with the comparative advantage of cheap labour in low-income countries.

The second stage sees the rapid absorption of surplus labour, which pushes the economy towards its turning point, where the labour market tightens, signalling the need to shift to higher productivity employment. This heralds the third stage, which sees entry into sectors of emerging competitive advantage in HCI, such as steel and engineering, and the gradual abandonment of low-productivity manufacturing sectors, like textiles, to the next wave of NICs. This process may be accelerated either as an HCI drive or an HCI big push. The HCI big push is the more ambitious option: it seeks to capture the economies of scale by simultaneously establishing complementary intermediate and downstream activities, so that both sets of factories may operate at a larger scale and therefore capture the economies of scale (Murphy *et al.* 1989).

The merits of an HCI big push are disputed (Yoo 1990). It typically results in a three-stage sequence of a construction boom, deflation, and economic rebound (Auty 1992). The construction stage of an HCI big push hits the economy like a commodity boom, as shortages of critical factors of production trigger inflation, causing the exchange rate to strengthen and fiscal and trade deficits to widen. This requires economic stabilization policies which, however, depress economic growth below its long-term trend just as the HCI projects come on stream. This results in low HCI capacity use and an inadequate return on capital. But, provided economic stabilization is successful, there is likely to be an HCI-led economic rebound some four years after the onset of the downswing as the recovery in demand permits the new HCI plants to operate at full capacity and to generate an adequate cash flow for expansion.

Whatever the merits of the differing modes of diversification into HCI (and the controversy is examined in more detail with reference to South Korea and Taiwan, in Chapter 11), the third stage of the East Asian development model is associated with rapidly rising real incomes. This process continues in the fourth stage of the model, in which the economy is liberalized as it nears developed country status and becomes too complex for previous forms of state intervention to be effective. Meanwhile, investment shifts increasingly towards knowledge-intensive products such as new materials, information technology and pharmaceuticals. Table 5.1 summarizes the relationship between industrial policy, macroeconomic policy and economic performance.

In the context of the East Asian development model, both the large market-rich, low-income Asian countries and the large market-rich and resource-rich Latin American countries skipped the stage of labour-intensive exports. Similarly, many mineral economies and other resource-rich but market-poor primary product exporters in Latin America and also in sub-Saharan Africa used the rents from their primary sectors to pursue ambitious ISI strategies. They frequently entered new sectors well ahead of achieving an adequate domestic demand. Such forced industrialization invariably proved premature and mistaken.

In the case of the large low-income Asian countries, the policy of HCI-led growth reflected the priority which their governments gave to economic self-sufficiency. Both India and China provide clear evidence that, whatever the strategic justification for such a policy, HCI-led growth demands a high level of investment and yields a relatively slow rate of economic growth. Such a rate of investment calls for a considerable sacrifice in terms of consumption foregone by the poor rural majority. Yet HCI-led growth has also proved extremely difficult to reform, as the example of both India and China show.

Indian industrial policy and performance

India's HCI big push

The Indian planners launched an HCI big push within the second five year plan (1956–61). The Indian big push was initially successful in three respects. First, it achieved a marked increase in domestic saving and investment. The domestic saving rate rose from 10.4 per cent of GDP under the first (1951–55) five year plan, to a respectable 15 per cent in the early 1960s. Foreign capital

Table 5.2 *India: GDP and sectoral growth rates, 1951–89*

	Agriculture	Industry	Services	GDP
1951–60	3.2	6.3	4.3	4.1
1961–70	2.3	5.4	4.6	3.7
1971–80	1.5	4.0	4.5	3.1
1981–89	3.5	6.8	6.4	5.5

Sources: Mundle (1990) for 1951–1980, World Bank (1991) for 1981–9

helped lift the gross rate of domestic investment to 18 per cent of GDP by the mid-1960s. Second, the Indian HCI big push also succeeded in establishing a wide range of industrial sectors and their associated technologies (Bruton 1989). Third, the big push accelerated the growth rate of industrial production, from 4.7 per cent per annum during 1947–51 and 5.6 per cent through the first five year plan to 7 per cent in the second plan and more than 9 per cent in the third.

These gains were not sustained, however: the acceleration in economic growth slowed, technology lagged best practice, and the efficiency of investment declined markedly. The Indian GDP growth rate slowed from an average of 4.1 per cent through the 1950s to 3.7 per cent in the 1960s and 3.1 per cent in the 1970s (Table 5.2). But investment rose to 20 per cent of GDP by 1980, so that the slowdown in Indian GDP growth was accompanied by a substantial decline in the efficiency of investment.

Consistent with the three-stage big push model, the Indian HCI big push construction boom triggered fiscal and trade deficits which required deflationary measures in 1965 that heralded the second stage of the big push sequence. The economic deterioration was exacerbated by two external factors, namely the crop failure noted in Chapter 4 and war with Pakistan (which resulted in doubled defence spending, to 3.4 per cent of GDP, through the 1960s). The mid-1960s industrial slowdown was most pronounced in heavy metal-based capital goods (Ahluwalia 1985), where capacity use fell from 86 per cent in the third five-year plan, to 60 per cent in the early-1970s (Srinivasamurthy 1981, Ahluwalia 1985).

But there was no third-stage HCI-led rebound: the underlying growth rate of the industrial sector was barely 4 per cent per annum over the next 15 years – half the rate during the big push (Sandesara 1986). Ahluwalia (1985) blames the post-1965 economic slowdown on three factors: a deceleration in investment growth (especially in the public sector and in infrastructure), weak domestic demand for industrial goods, and an excessively protectionist industrial policy. The failure to abandon the autarkic industrial policy, despite the lip-service paid to helping agriculture, explains why Indian stabilization did not bring an HCI-led rebound.

Table 5.3 *India: absorption and structural change (% of GDP)*

	1950	1960	1970	1980	1985
Per capita GNP ($ 1980)	129	155	204	230	300
Absorption					
Private consumption	96.9	79.7	74.1	70.4	74.7
Public consumption	5.9	7.2	9.4	10.2	13.1
Gross investment: private	7.8	7.3	9.7	10.7	12.8
public	2.3	7.0	5.9	9.1	12.1
Value added					
Agriculture	62.2	50.1	47.4	38.0	31.4
Mining	0.7	1.0	1.0	1.6	3.0
Manufacturing	15.3	14.2	14.2	17.7	16.6
Light industry	11.9	7.8	6.5	8.2	5.3[a]
Heavy industry	4.2	6.4	7.7	9.5	10.9[a]
Construction	4.1	4.6	5.3	5.0	5.4
Utilities	0.3	0.6	1.1	1.7	2.2
Services	17.4	29.5	31.0	36.0	41.4
HCI value added					
Chemicals and rubber	0.8	1.8	2.4	2.8	3.1[a]
Non-metallic minerals	0.5	0.6	0.5	0.9	0.8
Basic metals	1.2	1.4	1.4	1.4	2.0
Machinery, engineering	1.7	2.6	3.4	4.4	4.6

Sources: World Bank (1989c) and CSO (various years)
[a]1984

Persistence and consequences of HCI dominance

The economic dominance of HCI persisted long after India's big push had officially been downgraded in favour of agriculture. Yet although India's GDP growth rate and efficiency of investment both finally improved somewhat in the 1980s (investment reached 25 per cent of GDP, while GDP growth averaged 5 per cent and restored the economy-wide ICOR from more than 6 to a still very modest 5), chronic trade and fiscal deficits persisted which eventually undermined the economic improvement.

The cause of these difficulties can be illustrated with reference to the changing production structure of the economy. Structural change in India departed considerably from the Chenery and Syrquin norms for a country of its size and level of development (Table 5.3). HCI expanded much faster than the norm, and by the 1960s it had outstripped light industry in importance when the country's per capita GNP was still less than $200 (in constant 1980 dollars). According to the norms, HCI should still lag light industry at $1000 per capita income (Wood 1986). Yet although the share of agriculture in GDP halved between 1950 and 1985, the manufacturing sector failed to make the

Table 5.4 *Comparative cold roll coil production costs ($/tonne)*

	India[a]	Mexico	Brazil	Japan
Material	62.0	82.0	40.0	61.0
Energy	134.0	77.0	130.0	86.0
Labour	21.0	60.0	50.0	65.0
Other	95.0	85.0	50.0	51.0
Total variable cost	312.0	304.0	270.0	273.0
Fixed cost	220.0	121.0	112.0	88.0
Total cost/tonne	532.0	425.0	382.0	361.0
Memo items				
Labour ($/MH)	1.0			12.30
Electricity ($kWh)	0.65			0.09
Scrap ($/tonne)	145.0			100.00
Coke ($/tonne)	43.0			67.50
Iron ore ($/tonne of Fe)	17.0			32.50
Depreciation and interest	46.2			
Levies	95.3			

Source: World Bank
[a]Indian figures for Rurkela, its highest-cost unit

expected compensation: it expanded its share of GDP by little more than 1 per cent (Table 5.3). This is because the early expansion of HCI was offset by a decline in the share of light industry in GDP – an unusual feature for a developing country.

Indian HCI maturation rates were well above the five to eight years needed to assure that the benefits from HCI compensate for the initial costs of infant industry protection. For example, the steel sector, which dominated the big push, far from maturing in terms of achieving international competitiveness and technical competence, actually regressed from the mid-1970s. This reflected the inefficiency associated with public sector ownership, inadequate infrastructure, and above all insulation from international competition. By the 1980s, India ranked among the world's highest-cost steel producers (Table 5.4), and quality was inferior to imported steel.

The use of HCI to promote regional policy did not improve its efficiency. The spatial dispersal of Indian HCI via suboptimal plants in remote locations dissipated the economies of scale and lowered economic efficiency, without narrowing regional income differentials. For example, the local multipliers from India's Deccan steel plants were weak (Farmer 1983), and the downstream growth of engineering accrued to a few established industrial cities (Rao 1984). Regional income differentials therefore persisted, as noted in the previous chapter, although some decentralization in industrial production was achieved (Saha 1990).

India's overly protected and slow-maturing manufacturing sector relied heavily on non-tariff controls (import bans) to keep at bay more efficiently produced and higher-quality foreign manufactured goods. Imports of all but a few essential consumer goods were curbed, and the bans imposed on capital and intermediate goods imports were applied in an *ad hoc* way. The average level of effective protection (i.e. protection expressed in relation to domestic value added) was in excess of 200 per cent (Lal 1988).

The average figure masks the fact that the trade regime discriminated against the more labour-intensive activities like agriculture and light industry, in which the Indian comparative advantage lay. Around two-fifths of Indian manufacturing enjoyed very high rates of protection (in excess of 70 per cent nominal protection, i.e. when expressed as a percentage of the total value of the good including purchased inputs as well as value added). These tended to be the more capital- and energy-intensive HCI sectors, while the 50 per cent of manufacturing with relatively low nominal protection (less than 30 per cent) has been labour-intensive.

In addition to the bias against light manufacturing, there has been a bias against exports: Indian export incentives have rarely compensated producers for the higher margins on sales to the domestic market. Not surprisingly, the slow-maturing HCI failed to generate exports commensurate with its importance in the economy. Even though HCI comprised almost two-thirds of manufacturing output (Table 5.3), light industry exports of gems, textiles and jute products made up more than two-thirds of Indian exports in the late-1980s. UNIDO (1985) concluded that, a generation after the launch of the big push, Indian comparative advantage remained in labour-intensive goods.

A more dynamic performance from the light manufacturing sector could have helped to cut the foreign exchange gaps during the 1955–65 HCI big push, and also eased subsequent foreign exchange crises, as well as absorbing surplus rural labour. But inadequate investment in labour-intensive light manufacturing such as food processing and textiles, which had been India's two main manufacturing sectors at independence, slowed their growth. The net effect was that the share of manufactured exports in total exports did not increase from the late-1960s to the mid-1980s. India's share of world exports halved from 2 per cent to 1 per cent in the 1950s, and halved again to 0.5 per cent by the early-1980s (Nayyar 1988). India also lags the trend for large economies to increase the share of trade in their GDP (Perkins and Syrquin 1989).

By the 1980s, the deep-seated foreign exchange and fiscal problems of HCI-led growth were proving counter-productive. They undermined India's self-reliance and jeopardized its reputation for fiscal prudence. Foreign debt doubled during the 1980s to 24 per cent of GDP, and debt service increased from a modest 10 per cent of exports to a barely supportable 25 per cent. Although some economic reform was triggered each time the foreign exchange constraint reached a critical level (which occurred in the mid-1970s, the mid-1980s and in 1991), the reforms that occurred were too mild.

The lack of dynamism within exports and light manufacturing slowed the absorption of rural labour (Deshpande 1986) and indefinitely postponed the labour market turning point. Since textiles and food processing accounted for 60 per cent of formal industrial employment in 1951, their slow growth was an important reason for the slow rise in total industrial employment. Even after the official downgrading of HCI in the mid-1960s, the growth of factory employment continued to disappoint (Dholakia 1986). In the mid-1980s manufacturing jobs accounted for barely 15.5 per cent of total employment (Sandesara 1986). This is well short of the level needed to alleviate poverty by easing pressure on the land and bringing the Indian labour market to the turning point.

In summary, India's HCI-led growth resulted in sizeable efficiency losses. It spawned a slow-maturing HCI sector which stunted rural purchasing power, attenuated the growth of competitive light manufacturing and entrenched a rigid high-cost command economy. It was associated with relatively slow economic growth, disappointing levels of investment efficiency and postponement of the labour market turning point. India's unfortunate experience parallels that of China, but the economic crisis triggered by the HCI big push occurred sooner in China (in the late-1950s) and reform, when it came, proceeded much faster. But the pace of the post-1978 Chinese reforms triggered unexpected problems, including tensions between the central government and the regional governments, and also between the more dynamic coastal regions and the lagging interior regions.

China's development strategy

China's big pushes

There have been three HCI big pushes in postwar China. The first two were planned, and comprise the Soviet-inspired big push of the mid-1950s and the defence-driven big push launched in the late 1960s. The third HCI big push was not planned, and occurred during 1984–8 as an unintended by-product of the application of the Dengist reforms to the manufacturing sector. During the initial big push of the first five year plan (1953–57) China's GNP grew very rapidly (Table 5.5). However, the first HCI big push ended in the disaster of the Great Leap Forward, which triggered a massive contraction of the economy between 1958 and 1961.

In a command socialist economy like that of China, the effects of an HCI big push construction boom (inflation, and expanding fiscal and trade gaps) would be suppressed. Rather, the impact would be transmitted in terms of the diversion of capital and human resources to HCI, with corresponding withdrawals from consumption-oriented activities such as light manufacturing and agriculture. The result would be shortages of both consumer goods and

Table 5.5 *Sources of economic growth in China (%contribution to net material product)*

Period	1953–7	1958–66	1966–76	1977–85
Capital stock	0.8	1.9	2.8	3.3
Labour force	1.7	1.6	1.8	1.7
Productivity	3.7	(1.3)	0.6	4.1
Total	6.2	2.2	5.2	9.1

Source: Economist (1987, p. 18), after Perkins (unpublished)

HCI goods. Such shortages did emerge in China by 1957, and they adversely affected agricultural production (Riskin 1988). The Great Leap Forward encouraged the establishment of large numbers of small communal 'backyard' HCI plants in the winter of 1957–8, in an attempt to make up for HCI shortfalls in the modern sector (Prybyla 1987).

The Great Leap Forward drew even more resources from agriculture, and worsened the already severe food shortages. The failure of the Great Leap Forward was followed by an economic slowdown: industrial output did not recover to pre-Leap levels until the mid-1960s (Bunge 1981, Cole 1988) and the average GDP growth rate during 1958–70, was 4.3 per cent, significantly down from that of the mid-1950s (Table 5.5). The deceleration in economic growth was the more disappointing because it occurred despite an increase in the rate of investment. By 1970, annual investment absorbed 29 per cent of China's GNP, twice the Chenery and Syrquin norm for a large low-income country, and a reflection of the powerful resource mobilization capacity of the command socialist economy.

Such a high level of investment, if deployed with Korean efficiency, could have propelled the Chinese economy at a 10 per cent annual growth rate. But China's GNP growth was barely half that under the fourth five-year plan during the remainder of the Cultural Revolution. Such low investment efficiency meant that, as in India, growth was driven overwhelmingly by increases in the labour force and in capital accumulation rather than by improved productivity.

The Chinese economy did surge briefly during 1969–73, however, partly reflecting the second HCI big push and partly recovery from the upheavals of the early years of the Cultural Revolution. With the second big push, the planners sought to establish HCI in the interior provinces, for reasons of military security. This decentralization of industry aimed to make each province self-sufficient, so that a military invasion of China would need to be total to be effective (Rothenberg 1987). The GNP growth rate averaged 11.8 per cent annually during 1969–73 (World Bank 1989c).

Overall, the growth of Chinese industrial output pushed the sector's share of value added from 15 per cent in 1957 to 45 per cent by 1976. But as in

India, HCI maintained its initial dominance despite attempts to downgrade it in the late-1950s and again from the late-1970s. The ratio of investment in HCI to that in light industry remained in excess of 8:1 into the 1980s (Prybyla 1987). Although prior to the Dengist reforms the productivity of industrial investment appears to have been inferior to that of agriculture (while within industry, the return on HCI was lower than that of light manufacturing), HCI continued to dominate Chinese investment. Data for 1965 and 1977–9 show that in each time period, China allocated one-fifth of total investment for agriculture and 55 per cent for industry and infrastructure (four-fifths of that for HCI) (Riskin 1988).

The unsuccessful efforts to downgrade HCI under command socialism were part of a wider political struggle between Mao and the reformers, led first by Liu Shao Shi in the early-1960s and then by Deng Xiaoping in the early-1970s. Mao was disgraced following the Great Leap Forward, but he ousted the reformers during the Cultural Revolution, so that sustained economic reform was delayed until after his death in 1976. The Dengist reforms introduced in 1978 aimed sharply to improve the Chinese standard of living by boosting consumption and lowering investment, especially investment in HCI.

Investment efficiency

Despite the unusually high rate of investment during the 2 decades prior to the Dengist reforms, per capita consumption in China grew no faster than that of other low-income countries. The dominance of HCI during those years reduced Chinese consumption in two ways. First, the high Chinese rate of investment (30 per cent and more of GNP) entailed a corresponding withdrawal of resources from more consumption-oriented activity. Second, the resources invested in HCI generated a relatively low return. The ICOR (which, it may be recalled, measures the ratio of extra investment to additional output), is estimated at around 5 for the years 1957–79. This compares unfavourably with an ICOR of 3.9 for mid-income countries during 1960–78 (World Bank 1981).

Apart from the resource allocation inefficiencies inherent in the pursuit of industrial autarky, a further source of inefficiency lies in China's large state-owned enterprises (SOEs) which, as in India, dominated HCI investment and produced more than four-fifths of industry value added in 1979. A further cause of poor investment efficiency was China's regional decentralization policy. For example, compared with Shanghai, the seven western provinces possessed three times as much industrial capital but generated scarcely half as much profit and tax (Wanging Guo 1988).

Three decades of HCI-led spatial decentralization failed to eliminate the wide regional variations in Chinese levels of income, as Table 5.6 shows (Cole 1987, Li 1990, Zweig 1987). Prior to the reforms, the per capita income of

Table 5.6 *Chinese per capita regional gross value of industry and agriculture (1980 prices)*

Value	1952 Yuan	1952 % National	1978 Yuan	1978 % National	1980 Yuan	1980 % National
East coast	184.5	133	847.9	144	982.2	141
Centre	118.5	85	469.4	80	537.4	80
West	92.1	66	312.3	53	391.4	56
National	139.2	100	589.6	100	694.5	100

Source: Wanging Guo (1988, p. 35)

rural areas declined to one-third the level in urban areas. However, China was more successful than India in providing basic needs: the average daily calorie intake was estimated at 2500 during 1980–2, compared with 1900 for India (Perkins and Yusuf 1984).

But HCI-led growth was not sustainable. The inability of the neglected light industry sector to absorb surplus rural labour was reflected in a marked fall in rural labour productivity between 1957 and 1975. The sown area expanded by only 0.2 per cent annually 1952–79, while the workforce increased by 2 per cent annually, so that the sown area/worker fell to two-thirds of the 1952 level. Output rose because more workers were absorbed and because those in work increased their annual days, worked from 170 in 1957 to around 250 in 1975.

Total employment in agriculture grew from 231 million to 329 million during 1957–75, while manufacturing employment only expanded from 15 million to 40 million (Rawski 1979). Even so, the share of agriculture in total employment remained constant and in the early-1980s resembled that of India, at around 70 per cent (World Bank 1985a). The 1957–75 agricultural trend of ever-increasing labour inputs to offset declining labour productivity was not sustainable.

Constraints on Chinese industrial reform

Local government intervention

The Dengist reforms aim to raise real per capita income (in 1980 dollars) from \$300 to \$830 between 1981 and 2000. They seek to shift resources to consumption and to redeploy investment towards agriculture, light industry and infrastructure. A key component of the reforms is the decentralization of decision-making to the production units (farm or factory) in an attempt to raise the efficiency of resource use. In addition, economy-wide improvements

Table 5.7 *Comparative production structure and Chinese projections, 1981 and 2000 (% GNP)*

Year	1981			2000		
Country	Large Low-income	Large Mid-income	China	China Quadruple	China Modernize	China Balance
Agriculture	35	18	40	30	32	29
Manufacturing	17	29	43	48	49	45
Heavy industry	n.a.	n.a.	26	30	31	28
Infrastructure	10	14	10	14	13	14
Services	38	39	7	7	7	13

Source: World Bank (1985b, pp. 23 and 42)

in productivity call for the rejection of autarky, both regionally and nationally. The reforms seek increased specialization, which in turn requires an expansion of trade and of the national transport system.

The World Bank considers that if HCI is downgraded (the 'quadruple' strategy in Table 5.7) in favour of farming, light manufacturing and services (the 'balance' strategy in Table 5.7), then less investment will be required to achieve a given rate of economic growth. The switch away from HCI enhances the prospects of reaching the targeted 5.5 per cent per capita annual growth rate. It should also improve the efficiency of investment and accelerate employment creation, with the service sector becoming a more significant component in the Chinese economy.

The initial reform focused on agriculture: it took place from 1978 to 1984, and went well, especially after improved guarantees were given over security of land occupancy for family farmers. The decentralization of decision-making proved effective, because the family had already become the principal unit of economic activity in the rural economy (Hussain and Stern 1991). The rate of GDP growth increased rapidly during this period, even though the annual rate of investment declined to 30 per cent of GDP, as the Dengist reforms had projected. The reforming Chinese economy grew at 8.6 per cent annually under the fifth five-year plan (1976–80) and at 11.1 per cent during 1981–5 under the sixth plan.

The Dengist reforms initially stimulated farm production and shifted the industrial priority firmly from HCI towards textiles and consumer goods. Consumption increased its share of national income in the early years of the reforms, as the share of investment declined from 36 per cent to 30 per cent of GNP during 1978–83 (World Bank 1985a). Trade also increased as the economy opened up: exports exceeded 7 per cent of GDP in 1980, and although this is still two-thirds of the Chenery and Syrquin norm, that figure

Table 5.8 *China: structural change and absorption, 1952–85 (% GNP)*

Per capita GNP ($1980)	1952 175	1970 245	1980 290	1985 445
Absorption				
Private consumption		55.3	55.0	51.7
Public consumption		15.5	16.3	13.9
Investment		29.2	30.0	38.7
Total		100.0	101.3	104.3
Exports	2.1	2.8	7.1	9.9
Value Added				
Agriculture	63.0	35.0	32.0	31.2
Industry	8.5	41.1	48.0	45.3
Manufacturing		31.7	35.3	33.4
Construction and utilities		9.4	12.7	11.8
Services	28.5	24.0	20.0	23.4
Memo item: adjusted 1981 structure of production				
Agriculture			35.8	
Industry			37.9	
Mining			10.8	
Light manufacturing			8.1	
HCI			16.4	
Electricity			2.6	
Construction			4.7	
Services			17.2	

Sources: Perkins (1988), World Bank (1989c), World Bank (1985a)

compares with 2.8 per cent in 1970 and 2.1 per cent in 1952 (Table 5.8). Yet despite 3 decades of HCI-led growth, primary products accounted for almost half of all Chinese exports and manufactured exports were still dominated by textiles (one-fifth of total exports). This reflects the slow maturation of Chinese HCI which, like that of India, remained internationally uncompetitive.

The downgrading of HCI proved difficult for China to achieve. The early Dengist reversal of the dominance of HCI was ended unexpectedly when the reforms were applied to industrial production in 1984. This is because the devolution of decision-making to enterprises unintentionally triggered a third HCI big push. HCI's share of investment, having fallen from 49 per cent to 36 per cent between 1978 and 1985, climbed back to 44 per cent by 1988. The consumption gains of the early reforms were reversed as the contribution of investment to growth more than doubled, comparing 1985–8 with 1980–4. The growth in industrial output accelerated from 9.9 per cent during 1980–4 to 17.8 per cent during 1985–8 and jeopardized the entire reform strategy, as

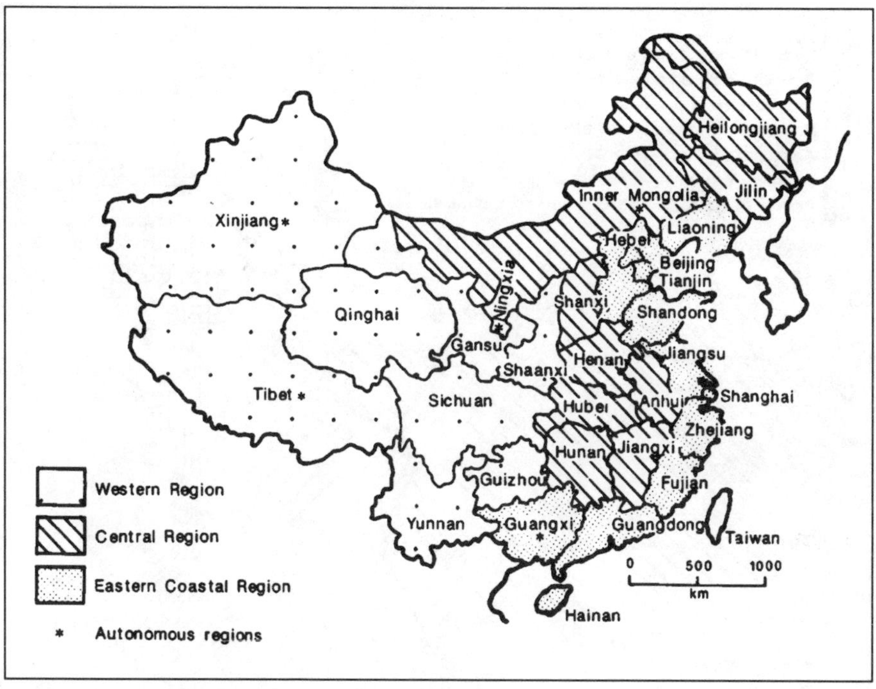

Fig. 5.1 *Chinese provinces and tripartite regionalization in the 1986–90 plan*

the prices of many goods rose. This depressed worker incentives: grain production stagnated during 1984–7, and urban living standards fell (Zweig 1987).

Regional imbalances

The industrial phase of the reforms aimed to decentralize many decisions down to the level of the manufacturing enterprise. But local bureaucrats intervened and captured more authority than was intended. As a result, the control ceded by the central government was not adequately compensated by increased flexibility in the economy. The attempt by local tiers of government to practise regional autarky is the principal cause of the unintended resurgence of HCI during 1984–8: it reflected a lengthy struggle over regional policy.

Chinese efforts to disperse economic activity had begun with the first five year plan, which had aimed to reduce the prominence of Shanghai (which, with the north east, had dominated industrial production before the Chinese

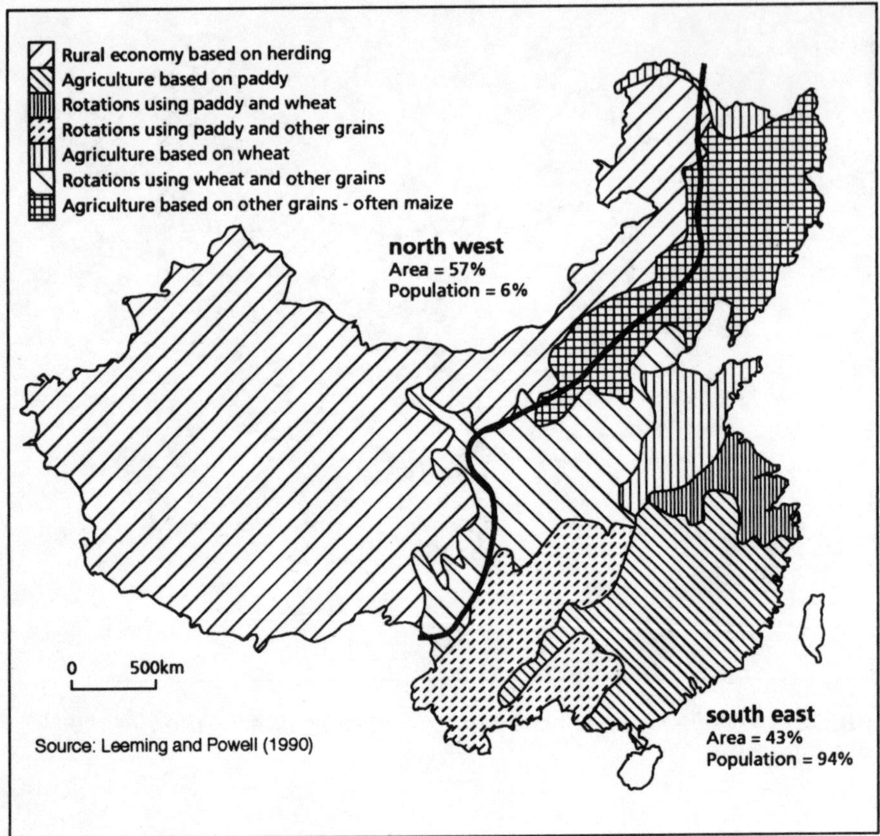

Fig. 5.2 *Major land-use patterns in China*

revolution). Major steel complexes were built at Wuhan (Hubei) and Baotou (Inner Mongolia), while the war-damaged factories of the north east (including the Anshan steelworks in Liaoning) were refurbished (Fig. 5.1). Yet 20 years later, the north-eastern third of the country still accounted for three-quarters of industrial production (Bunge 1981: 240).

For strategic reasons the 1969–73 HCI big push deliberately channelled resources towards backward interior locations and away from the coastal areas. Provincial self-sufficiency was fostered at the expense of economic efficiency. Consequently, the interior provinces generate a much higher fraction of gross value of industrial output from HCI than from light industry, irrespective of their natural resource endowment and local market size.

Hubei province provides an example of the inefficiency of local monopolies: one town has four steel units, each reporting to a different administrative level (national, provincial, municipal and county). Yet all plants except the national enterprise are of suboptimal scale, encounter material supply problems and perform poorly (World Bank 1985b). A more rational (and competitive) production system would close the smaller units.

The Dengist reforms aim to build on the varying endowments of different Chinese regions. Advanced regions should concentrate on increasingly sophisticated manufactured goods and import intermediate inputs not produced locally from the less advanced provinces of the centre, or from abroad. At the other extreme, the backward western region might expand ranching, farming, forestry and transport (Fig. 5.2). The lagging regions are expected to catch up through a combination of specialization gains and the operation of 'trickle down' effects (Rothenberg 1987). Capital resources will still be transferred to the interior provinces from the east, but they will accumulate as reserves until lagging regions can effectively absorb them (Wanging Guo 1988).

The 1978–84 rural reforms rendered surplus a large number of peasant workers and spurred the growth of rural industry in order to slow movement off the land (*Economist* 1987). Rural enterprises may have increased three-fold during the mid-1980s, and they are estimated to account for 120 million jobs. The expansion of high-autonomy town and village enterprises is most evident in light industry and in exports in the coastal zones. But rural workers still flocked to mid-sized and larger cities, so that the level of urbanization doubled to more than two-fifths of the population through the 1980s (Chen 1991). The industrial reforms heightened income differentials within the cities between state bureaucrats on fixed incomes and both the peasants entering the expanding urban service sector and the industrial workers receiving profit-related bonuses. The industrial workers began to feel threatened by migrant rural labour, as they faced the loss of job security in the reforming SOEs (World Bank 1990b).

Although inter-provincial income differentials appear to have initially held steady through the reforms, by the late-1980s the faster growth rates of the designated coastal enterprise zones became apparent. By 1988 the zones extended along the entire eastern coast to embrace one-fifth of China's population but only 3.3 per cent of the land area (Chen 1991). Within the coastal fringe, the spectacular growth of the Shanghai and Guandong regional economies pushed per capita incomes there to levels more than three times the national average. Economic growth in these regions was propelled by the expansion of light industry, much of it for export and using foreign investment.

There is little evidence, however, that this coastal activity has been a catalyst for technical change in the rest of China, where the stress remains on local HCI self-sufficiency. If the interior zones persist in the pursuit of inefficient local regional import substitution, the productivity of their investment will fall even further below that of the coast and widen the income gap between themselves and the coastal region. Closure of the gap requires greater reliance on market-generated signals to guide efficient investment.

One method of doing this would be to devolve ownership to firms (not necessarily private companies) which can pursue profit maximization (Wood 1990). Such changes will lead to a resumption of the downgrading of HCI, which began in earnest in 1978. The fraction of GDP available for consumption by the Chinese people should improve as a consequence.

Conclusion

Although an inward-oriented industrialization strategy has been especially attractive to large developing countries, it limits competition and encourages rent-seeking behaviour, resulting in the inefficient use of capital and labour. Increased self-sufficiency requires the rapid expansion of HCI, which is both capital- and skill-intensive. This is particularly damaging in a large low-income country because it is deflected from its comparative advantage in labour-intensive activity.

India shows that an inward-oriented policy is associated with slow economic growth, growing income inequality, low investment productivity and recurrent fiscal and trade deficits. At the sectoral level, these problems reflect slow HCI maturation, rural neglect and an attenuation of light industry. With a more market-conforming industrial policy such as that pursued by Korea or Taiwan (typically yielding an ICOR of 3), India's mid-1980s' investment of 25 per cent of GDP, and population growth of 2 per cent, would have allowed per capita income to double every 12 years—and the benefits would have been more equitably distributed than under HCI-led growth. Moreover, the industrial mix would have been less energy-intensive and polluting than HCI.

Although China's socialist command economy sustained even higher levels of investment for HCI than India's capitalist command system, it was initially similarly unsuccessful in making efficient use of such investment and in downgrading HCI once it gained dominance. Given the sacrifices in terms of consumption foregone which are implied by HCI's high capital requirements and neglect of employment-intensive agriculture and light industry, the initial emphasis of the Dengist reforms on rural change represented a prudent downgrading in the role of HCI. But China's planners had more difficulty in applying the reforms to industry and also in integrating the regional economies into an efficient national system. The consequences of inward-oriented industrialization and spatial imbalances are discussed further in Chapters 6 and 7 with particular reference to urbanization and Latin America.

PART 4

Primacy and hyper-urbanization: mid-income Latin America

6

The city size controversy: Latin American experience

This chapter focuses on the problems of the cities, using examples mainly from the most urbanized group of developing countries, Latin America. But first Latin American development must be set in the context of the policy options discussed in the previous chapter. This clarifies the relationship between urbanization and development strategy. The second section of the chapter then explores the growth of developing country cities in terms of the nature and consequences of rural–urban migration.

The chapter then assesses the pessimistic structuralist theses concerning marginalization and hyper-urbanization (which are based on the premise that urbanization has been 'too fast'). It notes that recent empirical studies query the earlier pessimistic views. This leads into a discussion of the optimum city size, a topic which is analysed with reference to Mexico City, the largest developing country city. Potential solutions to urban problems are reserved to Chapter 7.

Latin American development and urban primacy

After the 1930–45 phase of involuntary import substitution, Latin American countries drifted into inward-oriented development strategies, with disappointing consequences for their economic performance in the 1970s and 1980s. Latin America's two large market- and resource-rich countries, Mexico and Brazil, favoured especially high levels of autarky, but most countries built up large slow-maturing manufacturing sectors. Most Latin American countries maintained overvalued exchange rates, in contrast to the large cumulative depreciations resorted to by India and China to improve their competitiveness through the 1970s and 1980s (Wood 1988). The strengthening exchange rate sapped the competitiveness of the Latin American countries' primary sectors and also retarded their competitive industrial diversification. An overvalued exchange rate makes imports cheaper and can be used to justify the retention of protection for infant industries, thereby further postponing their maturation.

A recurrent Latin American response to the fiscal and foreign exchange deficits of inward-oriented industrialization has been to 'grow out' of these problems through accelerating public spending. Such 'populist' booms are meant to avoid the austerity associated with the initial stabilization stage of orthodox reform, but they are counter-productive (Dornbusch and Edwards, 1991, Sachs 1989). The booms have a typical 4-year cycle, which first brings a brief period of rapid economic growth as wage rises outpace inflation and boost demand, so that idle productive capacity is fully utilized. The boom accelerates inflation and creates even worse trade and fiscal deficits. Economic stabilization is required, whose deflationary measures cause a sharp economic contraction which often pushes incomes below pre-boom levels (Sachs 1989). The two most important Latin American countries, Mexico and Brazil, provide classic examples of self-defeating boom cycles.

Mexican postponement of industrial reform

In Mexico, ISI became entrenched during the country's pioneering green revolution of 1945–65 (Reynolds 1970) which (temporarily) transformed the country from a food importer into an exporter. But the Mexican green revolution ran out of steam by the late-1960s, as large farms grew short of land and smaller farms experienced difficulty in adopting the new techniques. The inability of the primary sector to continue to underpin industrialization is consistent with Mahon's (1992) observation about the role of primary product exports in Latin American development. The expansion of such exports at first raises per capita incomes but, as the primary sector shrinks in importance relative to manufacturing and services, the slow-maturing industrial sector cannot then generate adequate foreign exchange and tax revenues.

As the problems of autarkic development worsened, the Mexican government back-tracked on industrial policy reform, in favour of a public sector boom during 1972–5. The failure of this policy (Table 6.1) led to the espousal of outward-oriented reform in 1977. But a decision to accelerate the exploitation of the country's newly announced oil reserves led quickly to the abandonment of reform in favour of a growth-based solution via an HCI big push during 1978–82. In effect, Mexico used its oil resource to avoid the need to make its manufacturing internationally competitive.

Mexico briefly acquired some features of a mineral economy at this time (see Chapter 9). The 1978–82 big push boosted inflation and caused the exchange rate to strengthen (Table 6.1). This halted the country's modest non-oil export growth, and by 1982 virtually the entire non-oil economy was non-tradable, that is, in need of quota protection or subsidies. The Mexican use of the oil windfall for an HCI big push provides a clear-cut example of the resource curse thesis. Only in 1986, with the collapse of world oil prices, did the government finally accept that orthodox reform could no longer be

Table 6.1 *Mexican big pushes*

	Pre-push			Construction			Stabilization				
Echeverria big push	1970	1971	1972	1973	1974	1975	1976	1977			
GDP growth (%)	12.9	4.1	5.5	8.3	5.9	5.6	3.8	3.6			
Inflation (%)	10.9	5.9	6.0	13.2	22.1	15.6	19.5	29.9			
Real exchange rate (1965 = 100)	96.9	95.2	90.7	88.3	98.4	101.2	94.9	77.3			
Fiscal balance (%GDP)	(1.7)	(2.5)	(4.9)	(6.8)	(7.2)	(10.1)	(9.8)	(6.7)			
Current account (%GDP)	(2.9)	(2.0)	(2.0)	(2.6)	(4.3)	(4.8)	(3.6)	(1.9)			
Debt/GDP rate	15.9	15.7	15.2	16.4	17.7	18.5	24.4	32.0			
	Construction						Protracted stabilization				
Portillo big push	1978	1979	1980	1981	1982	1983	1984	1985	1986	1987	1988
GDP growth (%)	8.2	8.7	7.0	7.9	2.8	(5.0)	4.2	3.8	(4.9)	2.5	2.5
Inflation (%)	16.8	19.5	26.7	26.0	61.0	90.4	59.1	56.7	74.3	139.3	103.8
Real exchange rate	76.4	83.1	97.4	118.6	84.7	76.3	89.0	90.4	58.4	56.6	71.8
Fiscal balance (% GDP)	(6.7)	(7.4)	(7.9)	(14.7)	(17.6)	(8.9)	(8.7)	(10.0)	(16.3)	(9.9)	(9.6)
Current account (% GDP)	(3.0)	(4.4)	(6.6)	(7.4)	(3.0)	(3.1)	(2.6)	0.7	(1.1)	2.3	(2.0)
Debt/GDP rate	33.7	34.2	35.1	36.2	41.4	53.9	37.9	56.6	66.6	62.5	68.6

Source: World Bank (1991), Wood (1988), Beristain and Trigueros (1990).

Table 6.2 *The HCI big push sequences, Brazil*

Stage	Pre-push	1 Construction	2 Stabilization	3 Rebound
Brazil late-1950s push				
Time period	1951–5	1956–60	1961–7	1968–3
GDP growth (%/yr)	6.7	8.1	4.6	11.2
Inflation (%/yr)	16.1	21.8	53.7	21.0
Fiscal gap (%/GDP)	n.a.	(2.4)	(3.5)	(0.5)
Current account gap (%/GDP)	1.3	(1.1)	(0.3)	(2.2)
Brazil late-1970s push				
Time period	1968–73	1974–9	1980–4	1985–8
GDP growth (%/yr)	11.2	5.5	1.5	4.7
Inflation (%/yr)	21.0	41.4	132.1	317.8
Fiscal gap (%/GDP	(0.5)	(1.5)	(3.0)	(12.4)
Current account gap (%/GDP)	(2.2)	(4.9)	(3.8)	(0.3)

Source: World Bank (1991), UN (1970, 1976), IBGE (1987)

avoided. Autarkic industrialization was then abandoned rapidly, with little evidence of 'cultural' obstacles (Loser and Kalter 1992).

Brazilian HCI's protracted maturation

Brazil also used its rich resource base to sustain an autarkic industrial strategy and it persisted with autarky for longer even than Mexico. The country's first HCI big push occurred during 1956–60 and was based on steel and automobiles. It outstripped domestic implementation capacity and triggered the typical macroeconomic sequence of economic boom, stabilization and HCI rebound (Table 6.2). But stabilization was delayed by weak government during 1960–3, and only began under the military government, during 1964–7. A partial opening of the economy during 1967–73 then led to the 'economic miracle', when the economy grew in excess of 10 per cent annually. In part, the economic miracle reflects the delayed HCI rebound as HCI plants reached their design capacity.

Even with partial reform, however, Brazilian industrial protection was still relatively high, rents remained substantial (Bergsman 1970), income inequality continued to grow (Baer 1989, Simonsen 1989) and HCI sectoral maturation rates were orders of magnitude longer than those for Korea and Taiwan.

Exports remained small in relation to GDP and were heavily resource-dependent and subsidized. The reforms were far from consolidated when the Brazilian government responded to the 1973 oil shock by drawing on the country's large natural resource endowment to turn inward again and launch a second HCI big push. Much investment went into capital-intensive resource-based metal and hydro projects in the Amazon, in a vain attempt to develop domestic energy substitutes and also exports, to ease the deterioration in the trade balance which was caused by higher oil costs. State firms played a key role in the HCI projects; as in India, their performance deteriorated as their investment grew.

Brazil's second HCI downswing began in 1979. Adjustment lagged the rate of change required (Bacha 1986), so that the economic stabilization phase was once again protracted, as it had been in the early-1960s (Table 6.2). Far from maturing, key HCI sectors like steel and car assembly regressed from international best practice in the 1980s. Significantly, Brazilian negotiators cited the size of the Brazilian economy, when bargaining with the IMF to reduce the onerousness of orthodox policy reforms. This carried the implication that, if necessary, Brazil was large enough and had a sufficiently diverse natural resource endowment to 'go it alone', that is to cut itself off almost entirely from the international economy. A decade after the end of the second HCI big push boom had reversed the 'economic miracle' reforms, Brazil still lacked the consensus needed to reform its discredited autarkic industrial policy.

Other Latin American countries also used their rich natural resource endowment to bolster ambitious ISI policies, despite the smaller size of their domestic markets (Table 6.3). Colombia is an interesting exception, where a formal balance within the government between the political left and right maintained a policy which was both more consistent and more open than typical of Latin America (Cohen and Gunter 1992). Elsewhere, as the shortcomings of inward-orientation emerged, most countries sought to avoid the austerity associated with orthodox reform. The scale of the consumption cuts to Latin America's relatively high per capita income that were required in order to embark upon export-led industrial growth seemed to be such that no government could survive politically long enough to reap the benefits (Mahon 1992).

The region's overly protected domestically oriented manufacturing was unable to export in order to earn sufficient foreign exchange to service the debt. Consequently, most Latin American countries underwent a severe economic adjustment during 1982–92 – the 'lost decade' of the debt crisis (whose causes and consequences are discussed in Chapter 11). Many economies stagnated at that time, as around 6 per cent and more of their GDP was diverted abroad annually to repay the debt and associated interest. Unlike low-income Asia, most Latin American countries lacked a large agricultural sector to ease the adjustment costs of economic reform. A majority of the Latin American population had already migrated to urban centres, many to

Table 6.3 *Economic and social indicators for leading mid-income Latin American countries*

Country	Area (million sq. km)	Population (million)	Cropland/hd (ha)	Population growth (%/yr)	Population in cities 1991 (%)	GDP growth 1980–91 (%/yr)	Per capita income ($1991)	Literacy (%)	Life expectancy (yrs)
Argentina	2.767	32.7	1.11	1.3	87	(0.4)	2790	95	71
Brazil	8.512	151.4	0.52	2.0	76	2.5	2940	81	66
Colombia	1.139	32.8	0.16	2.0	71	3.7	1260	87	69
Mexico	1.958	83.3	0.28	2.0	73	1.2	3030	87	70
Latin America and Caribbean	20.507	445.0	0.40	2.0	72	1.7	2390	84	68
East Asia	16.369	1667.0	0.11	1.6	52	7.7	650	76	68

Sources: World Bank (1993b), WRI (1992)

Note 1 Data on the six small Central American republics appear in Table 7.1

2 Data on the Latin American mineral economies appear in Table 9.1

Fig. 6.1 *Principal cities in Latin America 1990*

Table 6.4 *Levels of urbanization in developing countries*

	Per capita income ($1990)	Urban population % Total		Mean growth(%)		Population in capital city as % (1990)		Population in cities >1m as % (1990)	
		1965	1990	1965–8	1980–90	Urban	Total	Urban	Total
South Asia	330	18	26	3.9	3.9	8	2	34	9
Sub-Saharan Africa	340	14	29	5.8	5.9	32	9	29	9
East Asia	600	19	50	3.0	12.0	9	3	30	11
Mid East and N. Africa	1790	35	51	4.6	4.4	27	13	42	21
Latin American and Caribbean	2180	53	71	3.9	3.0	23	16	45	33

Source: World Bank (1992a)

one or two dominant cities. Industries which located in such large cities had maximum access to the all-important protected domestic market, so that ISI was an important cause of the skewed city size distribution in Latin America (Fig. 6.1).

Country group patterns of urbanization

Table 6.4 shows that the mid-income countries of Latin America had 71 per cent of their populations living in cities by 1990. This compares with only 26–8 per cent in the low-income countries of South Asia and sub-Saharan Africa. The two other mid-income regions in Table 6.4, East Asia and North Africa-Middle East, have around half their populations urbanized. The fast-growing economies of East Asia experienced an especially rapid rate of urbanization through the 1980s due largely to the migration triggered by the reforms in China (which dominates the regional figure).

Among the low-income countries, South Asia has maintained a relatively modest rate of urbanization, at a level two-thirds that of sub-Saharan Africa. Many South Asian cities were already large in absolute terms by the mid-1960s and the relatively depressed urban incomes discouraged an exodus off the land on the scale seen in sub-Saharan Africa (or reforming China). Despite the faster rate of urbanization in sub-Saharan Africa, rural inhabitants will still comprise a majority of the population of that region by the year 2000 (as in South Asia).

A UN report in the mid-1970s (Beier *et al.* 1976) optimistically suggested that the relatively low rate of urbanization in sub-Saharan Africa gave that region the maximum scope for effective planning to avoid the problems encountered by more urbanized areas. In fact, as Table 6.4 shows, the urban growth rate in sub-Saharan Africa persisted at a relatively high level. The urban bias of policies in post-independence sub-Saharan Africa created too many subsidized and low-productivity, but relatively well-paid, city jobs (see Chapters 2 and 3).

By 1990 the Latin American and Caribbean countries had reached the mature phase of the urban cycle, with almost three-quarters of their populations already urbanized. Future urban growth in that region will be driven by natural increase within the cities, rather than by in-migration. The rapid emergence of very large cities such as Mexico City and São Paulo has left a back-log of inadequate infrastructure and excessive pollution; the search for solutions has been complicated by the economic squeeze associated with the debt crisis.

Growth of very large cities

Table 6.4 shows that as the level of urbanization increases, the numbers living in very large cities rises disproportionately. For example, more than two-fifths

Table 6.5 *City size trends, 1950–2000 (millions)*[a]

	1950	1970	1990	2000 Projection
Tokyo	6.75	14.87	20.52	21.32
Mexico City	2.88	8.74	19.37	24.44
São Paulo	2.75	8.06	18.42	23.60
New York	12.34	16.19	15.65	16.10
Shanghai	10.26	11.41	12.55	14.69
Calcutta	4.45	6.91	11.83	15.94
Buenos Aires	5.13	8.31	11.58	13.05
Seoul	1.02	5.31	11.33	12.97
Greater Bombay	2.90	5.81	11.13	15.43
Rio de Janeiro	3.45	7.04	11.12	13.00

Source: Tolba *et al.* (1992)
[a] Largest cities ranked by actual size in 1990

of the urban population of the most urbanized regions live in cities with more than one million inhabitants compared with around 30 per cent in the low-income countries. As a consequence of such trends, some of the world's largest cities are now in the developing countries (Table 6.5). Between 1985 and 2000 the population of Mexico City is expected to rise from 17.3 million to 25.8 million, that of São Paulo from 15.9 million to 24 million, and that of Rio de Janeiro from 10.4 million to 13.3 million.

Calcutta, Bombay, Tehran, Shanghai, Delhi and Jakarta are all projected to have more than 13 million inhabitants by the end of the century, as the lagging regions catch up with Latin America (Tolba *et al.* 1992). The growth of such cities is consistent with the phenomenon of urban primacy in the developing countries, in which one or a handful of very large cities dominate the pattern of settlement size distribution (Berry and Garrison 1958).

Considerable controversy surrounds the phenomenon of third world urbanization. Some argue that the process has proceeded far too rapidly, so that urban migrants have exceeded the capacity of the industrial sector to create jobs. Advocates of this view, like the sociologist Hoselitz (1957) describe the process as 'hyper-urbanization'. It has created a large 'marginalized' class with minimal economic, social and political links to the favoured social minority (Eckstein 1977). The position can be summarized in the case of Latin America by the 70/30 ratio, whereby around 70 per cent of the population remain in poverty while 30 per cent enjoy a relatively high standard of living. The Latin American ratio compares with one of 20/80 for the United States, where the 'underclass' of the poor comprises one-fifth of the population.

On the other side of the debate, economists like Richardson (1973) claim that the notion of hyper-urbanization is too extreme a depiction of the polarization of society. They argue that rapid urbanization reflects the orderly transfer of people from areas of low opportunity (the countryside) to areas of emerging high opportunity (the cities). The growth of very large cities in the developing countries in this view results from efficiency gains arising out of the internal and external economies of scale. This view has, however, also been disputed by environmental economists like Ward (1976). Ward has argued that the diseconomies of congestion outweigh any scale benefits from large city size. This view gained in strength through the 1980s, as evidence emerged of the scale of urban pollution in developing country cities (Hardoy *et al.* 1992). The city size controversy is examined below, with examples drawn mainly from Latin America, the most urbanized developing region.

Rural–urban migration

Evidence from the early postwar decades suggests that rural–urban migration was a step-wise process in which migrants moved up through the urban hierarchy, but with considerable slippage, or return migration. Initially, the migration may involve only the head of the household who moves a short distance. If the migrant is successful, the entire family is likely to move to the city. Migrants often move up through the urban hierarchy as a result of trial and error. For example, a survey showed that two-thirds of migrants to Delhi had tried between six and thirteen other cities before finally reaching the capital (Breese 1966). Such movement is consistent with the tendency for the very largest cities to grow faster than the smallest. For example, Taiwanese settlements of less than 50 000 population experienced a net loss of inhabitants, whereas those in the next largest size range (50 000 to 100 000) absorbed migrants at a rate of 5/1000, compared with 12/1000 for settlements of 100 000 to 200 000 and 16/1000 for cities with more than 200 000 inhabitants (Speare 1974).

Behind such figures there is, however, evidence of slippage which confirms the view that, in the early postwar decades, migration had involved trial and error with a very sizeable annual turnover. For instance, in the Taiwanese study, the ratio of in-migration to out-migration was 4 to 3 for the largest cities and 14 to 13 for small cities. The reverse flow evident in these figures reflects several factors. These factors include the initial short-term move of the household head, the difficulties encountered in coping with early setbacks, and also the return to the countryside of successful migrants, perhaps to inherit family land or to retire.

Causes of rural–urban migration

A lively controversy took place in the 1960s and 1970s over whether push factors (the hardships of rural life) or pull factors (the attractions of the city) accounted for rural–urban migration. The dispute was couched in terms of human capital models which assume that migration is a rational response to alternative employment opportunities. Migration entails costs, which include:

- the actual move itself (which is a function of distance);
- the earnings lost while the migrant finds and masters a new job;
- the psychic cost of the social disruption and adjustment to unfamiliar surroundings.

Such costs are set against the expected benefits, which include higher aggregate earnings at the new location for both the migrant and the other family members; and superior social services, entertainment and other experiences.

The most well-known migration model is the Harris–Todaro model (Harris and Todaro 1970). This model assumes that migration is a function of the difference between the urban wage and the rural wage and the probability of getting an urban job (which is itself a function of the level of urban unemployment). For example, if the urban wage is assumed to be twice the rural wage (a not untypical case), then the migrant may remain unemployed for up to 6 months in a year without being any worse off than if he or she had not migrated. In this way, the Harris–Todaro model postulates that the gap between urban and rural wages, together with the probability of finding a job, regulates the flow of migrants into the cities.

The analysis of human capital models of migration relies heavily on multiple regressions, the results of which have not always been very strong. One recurrent problem has been the low regression scores which suggest that the distance factor is a greater deterrent to migration than expected. For example, a study of Brazilian migration found that a move of 330 kilometres cost 75 cruzeiros, but migrants required 7 times as much in annual compensation to induce them to move that far (Levy and Wadycki 1974).

The relatively disappointing regression scores prompted some modellers to suggest that they reflected the migrant's imperfect knowledge of movement opportunities. In effect, the expected behaviour was being distorted by the presence of an intervening opportunity which presented migrants with a second-best, but less risky destination. Such an intervening opportunity might well be a smaller and closer settlement, one better known to the migrants than the 'more rational' larger distant city and one where the risks of migration seemed to be lower. The inclusion of variables to account for intervening opportunity can greatly improve the regression results, as Levy and Wadycki (1974) show for Venezuela. The notion of an intervening opportunity ties in with the slippage noted in migration behaviour.

Both the intervening opportunity model and the Harris–Todaro model support the primacy of pull factors over push ones in causing migration. Some researchers have queried the pull model, however, precisely because it makes unrealistic assumptions about the accuracy of the information available to migrants and the ability of migrants to make rational decisions. If the assumptions about migrant information are weakened, then migrants' behaviour becomes consistent with the observed step-by-step moves (trial and error) and slippage. A Turkish push migration study (Munro 1974) postulates that the propensity to leave a previous residence is higher:

- the greater the percentage of the origin population in towns of more than 25 000 people;
- the higher the level of literacy;
- the lower the wages in farming and the more limited the job opportunities in non-farm employment.

There is evidence that landless labourers may be more motivated to migrate by push factors than those living in farm families (Roy *et al.* 1992). Overall, it seems likely that push factors may set the migration process in motion during the very early stages of development. Pull factors then appear to take over as, through trial and error, the migrant learns more and his or her opportunities widen. However, as communications have improved, so migrants appear to make better-informed choices about their destinations, to move over longer distances and to reduce the degree of slippage (Skeldon 1990).

Consequences of migration

Other studies have investigated the consequences of migration, rather than its causes. Butterworth and Chance (1981) show that migration tends to be selective and that it drains away the more enterprising from the rural regions. For example, the 15–29 year old age group is most likely to migrate, with a 50 per cent greater probability than those aged 30–44, and twice the probability of those aged 44–59. In Latin America, women are more likely to migrate than men, especially in the 20–4 year old age group (Gilbert 1994). Such studies also show that migrants tend to be better educated than non-migrants and, consequently, that they are less likely to have just been employed on a farm than non-migrants. In Taiwan, for example, the children of wealthier farmers were most likely to migrate because the farmer could hire paid labour and his children therefore had more opportunity to try local non-farm jobs before seeking more distant non-farm employment.

The migration of the most enterprising in a village population raises doubts about the benefits and costs of such moves (Stillman 1973). Pessimists argue that the rural community expends resources on rearing migrants, but then loses its more able cohort to the cities before they have had much time to make a return contribution to the community which raised them. It is further argued that migration brings overcrowding to the cities, outstripping their capacity to provide jobs and shelter, while also increasing the incidence of alienation, social instability, crime, political unrest and misery.

Yet such pessimistic views ignore the fact that migrants do help rural areas in a number of ways. They reduce population pressure on the land, thereby improving the prospects for those who remain in rural areas of increasing farm size and incomes. Migrants often send money remittances back to home areas, and they visit their home community once or twice a year – or even return permanently. Such visits bring new ideas to the origin areas and enhance the knowledge of rural communities. In this context it is significant that the rate of migration is rarely highest from the very poorest regions.

Meanwhile, within the cities, migrants usually seem to feel and to be better off than they were in their previous home. For example, a study of Costa Rica shows that within 2–3 years most migrants are absorbed in higher-paying jobs and more steady employment (Caraval and Geithman 1974). More recent research suggests that social assistance for migrants when they arrive in the city has become substantial and that, as a consequence, over time migration has become less selective and has also tended to feed upon itself (Grindle 1988).

Summarizing, migration appears to be a rational reallocation of productive (human) resources from the lagging rural areas of low opportunity to urban areas of emerging high opportunity. Yet, while recognizing that such movements of people are a necessary part of the increased productivity needed to raise average incomes, many scholars have queried whether the actual speed and scale of third world urbanization is either necessary or desirable. Such views are reflected in the controversy over hyper-urbanization and marginalization.

Marginalization

The relationship of marginalization to hyper-urbanization

Hoselitz (1957) first proposed the hyper-urbanization thesis in the 1950s. He argued that urbanization in the mid-twentieth century, unlike the nineteenth-century process in the industrial countries, was outstripping industrialization. For example, by 1870 Sweden already had 22 per cent of its workforce in manufacturing industry, but only 11 per cent of its population was urbanized. A century later, Venezuela had only 9 per cent of its workforce in manufacturing, but 50 per cent of its population was urbanized. Hoselitz concluded that rural–urban migration was proceeding faster than the rate at which cities could provide adequate employment.

A concomitant of over-rapid urbanization is the marginalization of much of the urban population, which is excluded from participation in the modern, or formal, sector (Eckstein 1977). The latter provides secure employment at comparatively high wages for only a relatively small unionized 'worker aristocracy'. In contrast, the marginalized groups must scrape together a living in the informal sector while awaiting an opportunity to enter the modern sector. The employment options open to them comprise low-income artisan (craft) jobs and petty service jobs.

The jobs available to the marginalized groups are characterized as having ease of both entry and exit, because very little capital is required to set up in petty service provision. The intense competition in the informal sector eliminates any potential rents, leaving just a basic income to maintain a spartan urban life. Consequently, such workers have little ability to save and accumulate capital. In this way they fall behind those in the formal sector who, meanwhile, benefit from the presence of cheap informal labour through the purchase of cut-price goods and services. In addition, welfare benefits are absent or makeshift in the informal sector, because it pays no taxes and escapes the need to comply with social legislation. Such informal sector workers are assumed to reside in the run-down illegal squatter huts or inner city slums, and to have few social ties with the modern sector.

The marginalization thesis also embraces social and political relationships. For example, the marginalized groups do not constitute a threat to either workers or owners in the formal sector. Although wages in the informal sector are very low, they do not drive down wages in the formal sector in the manner of the reserve army of the unemployed in nineteenth-century Europe. In the modern sector, credentials are initially used to select workers (who are often over-qualified for the employment which they get) and the unions then maintain job security and pay, taking little interest in the welfare of the marginalized groups. Nor does the highly fragmented nature of the ownership in the small businesses in the informal sector create the cohesion required to pose a political threat to the formal sector.

Recapping, the marginalization thesis makes four important assumptions about the relationship between the formal and informal sectors. First, it assumes ease of entry into informal businesses, with little required by way of capital investment. Second, there is little opportunity to accumulate capital, because of intense competition. Third, there is minimal integration into the modern economy, so that the informal sector functions largely as a subsystem for the poor within the wider context of the formal capitalist sector. Finally, workers in the informal sector are thought to hold a strong desire to obtain employment in the modern sector, where unionization provides job security with relatively high wages and generous fringe benefits.

Conflicting evidence of marginalization

An influential study by Eckstein (1977) examined businesses in three areas within Mexico City. The areas included a city-centre slum, part of which dated from the colonial period; a suburban squatter settlement begun in 1954 and subsequently legalized; and an area of publicly financed low-cost suburban housing. Eckstein concluded from her analysis that local inhabitants were denied opportunities by the operation of international capital, and that any optimism over the employment prospects of the inhabitants of the squatter settlements was ill-founded.

Taking the example of artisan shoemaking in the city centre, Eckstein found that the typical enterprise had between three and ten employees. The owner handled the sales and knew all the skills, so that he could substitute for absent workers. The jobs were simple tasks such as leather cutting, sewing and glueing. The owners hired the cheapest labour – usually migrant women and children (despite the practice being illegal). The workers had no minimum wage, social security or redundancy pay, although the owner might help with medical fees, provided the workers went to a doctor friend (who charged a discounted fee). The sector was characterized by high rates of absenteeism, labour turnover and theft. Whereas the owners were expected to extend free credit to their customers, they were required to service loans from their suppliers promptly.

The 'modern' industry examined by Eckstein (1977) was near the city centre and comprised products such as textiles, food processing, paper bags and furniture. It operated in older premises with ancient equipment, mostly imported. The owners deliberately employed fewer than one hundred workers, in order to avoid legal obligation to pay fringe benefits. The employers, Eckstein found, were often the children of immigrants (especially from Eastern Europe), and discriminated against local residents by showing a preference for non-local labour, because it was more docile and reliable. They employed relatives in managerial posts and transferred capital out of the area, and were gradually being bought out by transnational corporations. These firms often relocated the operations to the suburbs, or retained the existing labour practices but introduced new machinery and cut the number of jobs.

Eckstein discovered little craft work in the two suburban locations she examined, although service premises were many and varied throughout all three regions studied. The shops ranged from a general corner store to specialist shops (like furniture), and there were many second-hand vendors. Although two-thirds of the businesses were locally owned, they tended to be small with an average age of 2 to 4 years, which reflected a high failure rate. Many of their owners had little business experience but wanted to be their own boss: small retail outlets offered them a route towards this objective through the investment of a capital windfall such as an inheritance or a lump-sum redundancy settlement. As with manufacturing in the city centre, most of the larger successful retail businesses were owned by foreign immigrants who relied on kinship links to get their firms started. Unlike many native Mexicans, they had the advantage of secondary education.

Eckstein's critique sharply contrasts the exploitative formal sector and exploited informal sector, but such studies have been challenged by subsequent work which suggests the situation is both more complex and less pessimistic (Butterworth and Chance 1981). For example, Peattie (1975) studied street vendors in central Bogota, Colombia, and found a much more sophisticated business environment than that portrayed by the marginalization theorists. Far from there being ease of entry into vending, new vendors were required to secure permits, which sold for up to one hundred times average daily earnings. Consequently, much capital was required to set up such businesses, as well as considerable skills in paperwork and bureaucratic manipulation. Moreover, the value of the permits was found to be positively correlated with site accessibility: the more central the site and the higher the pedestrian flow, the more expensive the site permit.

Nor did the informal businesses examined by Peattie have weak links with the formal sector. The goods they sold were not the products of workshops and peasant farms in the informal economy, but came instead from modern factories or urban food wholesalers. Nor were wages low: all vendors interviewed by Peattie made at least the minimum wage, and some fifteen times as much. They were not working in the informal sector in anticipation of eventually entering the formal sector. Rather, many valued their freedom and preferred to be their own boss than take a factory job. Far from being exploited, such operators appeared to thrive and to form a vital seedbed for entrepreneurial talent. As will be shown in Chapter 7 (which reviews solutions to the problems of large developing country metropolises), low-income housing also fails to conform to the expectations of marginalization theory.

Hyper-urbanization and the city size controversy

Overall, the urban population in developing countries grew by 5.5 per cent annually during 1960–73 and the population of the largest cities was typically doubling every fifteen years. In contrast, the nineteenth-century cities could not maintain themselves without in-migration (Hoselitz 1957). In 1841 southern Britain's life expectancy was 46, compared with 36 in London and only 26 in Manchester. Even in 1920, the urban death rate in Britain was still one-third above that of rural areas. In contrast to this situation, natural population increase is responsible for 50 per cent of the urban growth in South Asia and 70 per cent in Mexico (Davis 1965).

The aggregate numbers involved are certainly impressive: it has been estimated that developing country cities will need to have absorbed an extra 1.35 billion people between 1975 and 2000, pushing the number of third world urban dwellers to more than 2.15 billion (Beier *et al.* 1976). By contrast, over the same period the urban population of the industrial countries has been estimated to rise by one-third, to just over 1 billion. Meanwhile, the rural population in the developing countries will also have increased slightly, from 2.08 billion to more than 2.94 billion.

Causes of developing country city growth

Kelley and Williamson (1984) use data from forty countries over the period 1960–73 to model the causes of developing country city growth. Over that period the countries in their sample raised their GDP by 5.8 per cent annually, on an accelerating trend. The pace of urban growth rose from 4.7 per cent in the early 1960s, to 5.5 per cent in the early-1970s. But Kelley and Williamson argue that such a rate of urbanization is not unusually high, citing the experience of the United States in the nineteenth century, where the urban growth rate then averaged 4.8 per cent at a time when its population was growing by 2.6 per cent annually.

Kelley and Williamson analysed the sensitivity of their urban growth simulation to six 'drive factors': the rate of natural population increase, land scarcity, foreign capital availability, cheap minerals and energy, technological change and, finally, the rural–urban balance in technology diffusion. They found that neither population growth nor land scarcity had much impact on the rate of urbanization. For example, even if they held the supply of arable land constant (as opposed to an actual increase in land availability during the period they modelled), there was little impact on the rate of urban growth. Turning to the role of foreign capital, if such transfers are assumed to have been zero during 1960–73 (as opposed to an actual rate of 3 per cent of GDP), the urban growth rate is little changed. There is, therefore, no evidence to support Lewis's (1977) contention that the availability of large flows of capital from the industrialized countries had favoured a faster rate of urbanization in the late twentieth century compared with the nineteenth century.

Cheap raw materials – whether minerals or energy – did, however, display a modest positive correlation with increasing rates of urbanization. For example, falling mineral and/or fuel prices encouraged faster urban growth. But the key explanation for urban growth was technological change. The slowdown of industrial country productivity growth after 1973 led to a deceleration in the pace of third world urbanization. Closer examination of technological change revealed that it favoured urban-based activity rather than rural activity. This led Kelley and Williamson to conclude that the rate of third world urbanization is largely determined by the diffusion of technological advance from the industrial countries, a technology which has been predominantly urban-biased.

Kelley and Williamson tested their conclusions on the period 1973–9, when the actual rate of urbanization slowed to 4.3 per cent. This compares with a rate of 6.2 per cent, which their model forecast on the basis of the pre-1973 conditions. Kelley and Williamson attributed the deceleration to the higher costs of fuel and minerals, associated with the first oil shock, and also to the deceleration in technical advance. Looking to the future, Kelley and Williamson forecast that, for the developing countries as a whole, there would be a rapid fall in the urban growth rate between 1980 and the year 2000, to 2.6 per cent. By then, the fraction of the population living in cities would level off at around 80 per cent. Their projection, however, reckoned without the dramatic effect of the Dengist reforms in China, which almost doubled that country's level of urbanization within a decade.

The scepticism of Kelley and Williamson concerning hyper-urbanization is shared by many economists. It is mainly the sociologists who argue the counter-case. What does seem clear is that the choice of industrial strategy will affect the ability of the formal sector of the economy to absorb labour. In particular, the ISI strategies favoured by the countries of Latin America, sub-Saharan Africa and (pre-reform) low-income Asia retard the rate of rural labour absorption. This leaves a much larger fraction of the population for much longer than is necessary in a 'marginalized' informal sector (albeit not as oppressive as the more pessimistic observers have assumed). Such an outcome is consistent with the high levels of income inequality observed in Latin America and sub-Saharan Africa and the growing levels of income inequality under HCI-led growth in low-income Asia.

Moreover, the preference of manufacturers under ISI to locate close to the major market is consistent with urban primacy. Whatever the initial merits of the case for primate cities, doubts about the desirability of very large developing country cities have been reinforced by growing evidence of environmental deterioration.

Agglomeration economies and city size

The basic economic production function assumes that output is a function of capital and labour. A third variable can be added to capture locational effects, of which the agglomeration economies (i.e. the lower production costs arising from the concentration of economic activity in a single location) are the most important. City size can be used as a proxy for agglomeration economies, so that we can restate the basic production function to say that output is a function of capital, labour and city size. Two critical questions then arise. First, because a unit of labour or of land (part of capital) is often more expensive the larger the city, is this higher cost more than compensated for by the higher productivity with which it can be used in a large city? Second, if the unit cost of infrastructure rises with increasing city size, is this extra cost more than offset by the higher incomes and superior quality of infrastructure in large cities?

Empirical testing of these questions is difficult because so many variables are involved and much 'background noise' must be eliminated in order to isolate the city size factor. For example, the quality of labour varies with city size, in as much as larger cities tend to have a better trained labour force than smaller cities. A second cause of 'background noise' arises from the fact that larger cities tend to have a more capital-intensive mix of industrial sectors than smaller ones. Consequently, larger cities may have higher levels of productivity because they have larger shares of skilled workers working with more advanced equipment.

There are ways of muting such background noise by, for example, standardizing the weighting of different sectors among city sizes and then applying the actual capital and labour productivity figures for cities of differing size. By such means, some researchers have sought to measure how city size and productivity are related.

Studies for the United Sates suggest that, after standardization, large cities retain a productivity advantage. Sviekauskas (1975) observed a 5 per cent rise in productivity for every doubling of city size above a threshold of 50 000 population for eleven out of fourteen industrial sectors which he analysed. In another study, Segal (1976) found constant returns to scale for large and small cities, but large cities (over 2 million population) had higher productivity. In a third analysis, Kwashima (1975) found a steady increase in productivity up to 5 million, and then a decrease, suggesting a population threshold above which diseconomies of scale set in.

More recently, Henderson (1988) has analysed cities in Brazil and China and found that localization economies (savings which result from the clustering of similar industries) are more important than urbanization economies (savings from agglomeration enjoyed by most economic activity), and that localization economies are stronger for HCI than for light manufacturing. Finally, the localization economies fade with increasing city size. Overall, such evidence does suggest that the productivity of both capital and labour *will* be higher in larger cities, but that there are likely to be few extra benefits in *very* large cities. Richardson (1993) concludes that while the economic advantages for manufacturing in very large cities may be weak, policy biases which favour serving the domestic market and also securing the ear of those formulating policy, still favour location there.

Turning to infrastructure quality, the radical economist Edel (1972) has argued that while the private owners of capital capture the benefits of higher productivity arising from agglomeration economies, the community pays the cost in terms of diseconomies which manifest themselves in higher infrastructure charges and a less pleasant physical environment. But Richardson (1973) provided an interesting counter-argument, suggesting that in a free market, if labour did experience net diseconomies from living in large cities, workers would move to smaller centres to raise their real standard of living. Yet, as the discussion of migration suggests, this is not occurring and the net movement in population is up the settlement hierarchy. Moreover, migrants interviewed in large cities consider themselves to be better off.

Some interesting insight is thrown on the issue of social costs by Mera (1973), who notes that the relationship of infrastructure costs to city size tends to be U-shaped. This means that the unit cost of providing roads, electricity, sewage and education initially falls with increasing city size, and then subsequently rises. There appears to be a plateau of optimum settlement size of around 100 000 to 250 000 population. More detailed research suggests, however, that the quality of infrastructure improves markedly with city size, so that although costs are higher, the services provided are superior. Even more significantly, the real per capita cost of infrastructure falls with increasing city size. In other words, expenditure on service provision absorbs a smaller fraction of income as city size increases, so that the real cost of a superior infrastructure is less in larger cities.

In addition to the incidence of agglomeration economies and the real costs of infrastructure, a third major factor in evaluating large cities is the environmental cost. Although, as with the agglomeration benefits and infrastructure costs, much of the empirical research on urban environmental degradation has been undertaken in the industrial countries, the environmental problem is more acute in developing countries (World Bank 1992a).

Environmental pollution

The key urban environmental problems are water pollution, atmospheric pollution and traffic congestion. Measured in terms of dissolved oxygen, and taking 6.0 milligrams/litre as satisfactory (and higher levels as better), the quality of river water in high-income countries improved in the 1980s, from 9.5 to 9.9. The dissolved oxygen context figure held steady at 7.2 in mid-income countries, but fell from 7.2 to 6.4 in poor countries (World Bank 1992a).

Water supplies are often inadequate and deteriorating in many developing country cities. More than 1 billion people in the developing countries have no access to safe water, and some 1.7 billion have inadequate sanitation facilities. This means that the poor must often buy water from street vendors, which is typically 12 times more expensive than piped city water. For example, a poor Lima family uses one-sixth the water which a middle-class family uses, but its water bill is still 3 times higher. Such poor families may spend one-fifth of their income on water and women may spend 30–120 minutes daily obtaining water. These expensive and unreliable water supplies are also often of dubious quality, while sewage disposal creates serious problems, with excessive reliance upon on-site disposal systems which are smelly and not very hygienic.

Similarly, whereas air quality improved in the average developed country city to an adequate level of 50 micrograms/cubic metre of suspended particulates between the mid-1970s and the late-1980s, and also improved from 200 to 150 micrograms/cubic metre in mid-income country cities, in low-income cities air quality deteriorated from 300 to 350 micrograms/cubic metre. Pollution is much worse in present low-income developing country cities than in earlier ones at a similar level of development. Compared with WHO guidelines that suspended particulates should not breach a specific level on more than 7 days annually, studies in Beijing, Tehran, Calcutta and Bangkok reveal the level was breached on more than 200 days. Overall, around 1.3 billion people live in cities which fail to meet the WHO air quality levels, and the result is estimated to be 500 000 premature deaths annually and chronic coughs in 50 million children. In addition, high levels of lead emission from petrol have been associated with reduced IQs in children and high blood pressure (World Bank 1992a).

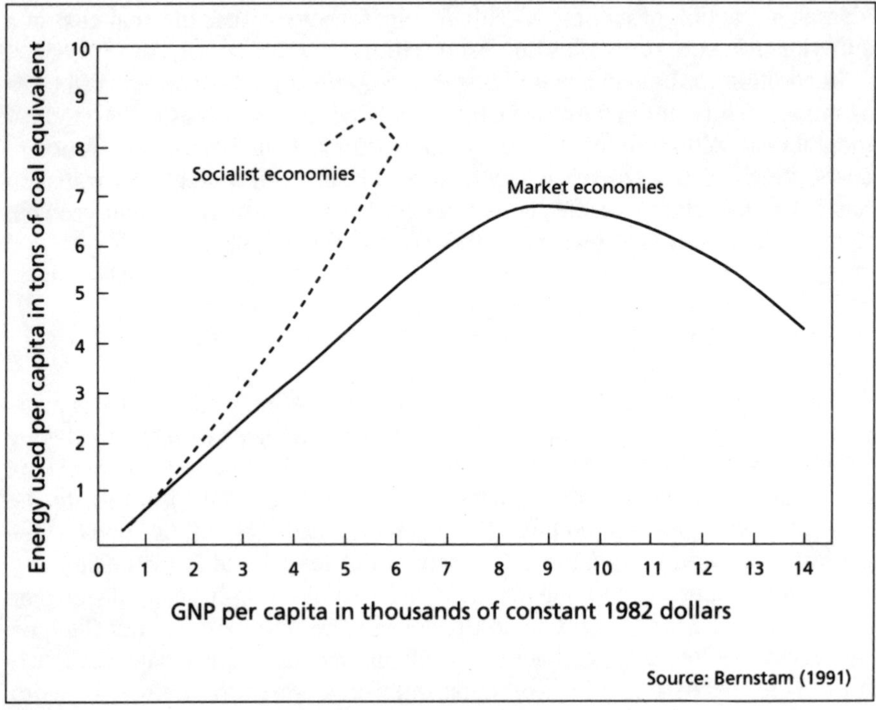

Fig. 6.2 *Per capita income and energy use*

This situation can be improved through appropriate policies, however. Bernstam (1991) estimates that both energy use and pollution emissions per $1000 of GDP show an inverted U-shape over time in market economies (Fig. 6.2). As per capita incomes increase, an initial rise in energy and pollutants gives way to a reduction in most pollutants. The socialist economies, however, show a continuing rise in energy consumption and pollution. These trends persisted even after their per capita incomes began to fall in the 1980s (Fig. 6.2). This pattern reflects the stress in the socialist economies on maximum output growth, irrespective of the cost in terms of either efficiency of resource consumption or the level of pollution emission. In contrast, the developed market economies have operated under a strong compulsion to use resources more sparingly and have curbed emissions at higher income levels, as consumers have displayed an increasing readiness to pay for environmental improvement.

The rapid decline in pollution at higher per capita income levels in market economies reflects a two-stage process: incentives are introduced to protect the environment and, then, cleaner and more efficient technologies diffuse in response to such incentives. For example, whereas the GDP of the OECD countries rose by 80 per cent during 1970–88, nitrogen oxides (emitted from vehicles) rose by only 12 per cent (and fell in Japan), sulphur oxides (associated with electricity generation) fell by 38 per cent, particulate emissions dropped by 80 per cent, and lead fell by between 50 per cent and 85 per cent. These achievements required the application of only a very small fraction of GDP (between 0.8 per cent and 1.5 per cent) annually to achieve. The fast-growing Japanese economy has set the standard because a larger fraction of its capital equipment has been newer and has embodied the latest technology. Such evidence led the World Bank (1992a) to conclude that it may eventually be possible for economic growth to be de-linked from environmental deterioration.

Congestion diseconomies: traffic and pollution in Mexico City

Mexico City rose to primacy during the present century. In 1900, Mexico City had a population of 350 000, only 2.5 per cent of the national population; 50 years later, however, the population had grown to 3 million, 12 per cent of the total. The share rose to 22 per cent by 1980, when Mexico City accounted for 40 per cent of the country's industrial activity (and manufacturing jobs), 30 per cent of its commercial employment and 40 per cent of the federal budget spending. Meanwhile, a shortage of space caused economic activity to spill over into the Central Valley of Mexico. That region alone comprised 12 per cent of the national land area, but housed 40 per cent of the population and provided 70 per cent of industrial output.

An oil producer, Mexico has set the domestic price of energy well below international levels. This policy has encouraged wasteful consumption and exacerbated problems of traffic congestion. In Mexico City, traffic congestion has slowed traffic flows to one-third the rate of Paris or London, despite the construction of expensive motorway and subway networks. The pollution problems of a vast metropolis are aggravated in the case of Mexico City by deficiencies in its site. A combination of anticyclonic weather, high altitude (2500 metres) and a valley location hamper windborne dispersal. The pollutants emitted by the city's 30 000 factories and 4 million vehicles are frequently trapped. Atmospheric conditions were so bad during the winter of 1990–1 that the ecology ministry twice declared an emergency, requiring industries to cut output by half. Some highly polluting plants were closed for good, including a $500 million oil refinery which had employed 5000 workers.

At the heart of the pollution problem lies the fact that the full costs of economic activity are not included in transactions, leaving the community as a whole to pay, in terms of a deteriorating environment. The Mexican government therefore launched a policy in 1991 (*Economist* 1991) which required investment by the state oil corporation in unleaded fuel and a switch to catalytic converters for new cars. But such measures will take at least a decade to be effective, because the car replacement rate is low (the average age of vehicles is estimated to be 11 years). Meanwhile, despite banning all drivers from using their cars on 1 day each week, the number of vehicles in the city continues to rise.

Further measures proposed for Mexico City include higher charges for city-centre parking, taxes on second cars (to block a popular middle-class loophole for evading the single-day car ban), and provision of cheap parking at suburban subway stations. Meanwhile, the World Bank (1992a) estimates that Mexico could halve vehicle emissions by simply imposing a tax on petrol to bring its price up to EC and Japanese levels. Such a tax would not only be socially cheaper than measures, such as catalytic converters, which aim to cut emissions per kilometre driven, it would also yield $300 million in revenues.

Returning to developing country cities as a whole, the World Bank estimates that the continuation of present practice would raise particulate emissions eleven-fold between 1990 and 2030. The removal of fuel subsidies (which would allow energy prices to rise to world levels) would give power stations the revenues to invest in improvements in the efficiency of resource use, while consumers would receive an incentive to conserve consumption. Even so, higher prices would still leave emissions 8 times higher by 2030 than 1990 levels. The addition of pollution controls (which in coal-fired plants can cut emissions by almost 100 per cent, for one-tenth of the cost of what is presently spent on subsidizing electricity) would leave pollutants below present levels by 2030, despite a very large increase in electricity consumption.

Conclusion

Like dependency theory, the hyper-urbanization and marginalization theses had their heyday in the 1970s and have since been challenged by statistical evidence. It has become increasingly clear that problems of rapid urbanization and low incomes are not so much an inherent consequence of capitalist development, as of the policy flaws which result from efforts to force the pace of industrialization. The legacy of such mistaken policies is at its most obvious in the most highly urbanized type of developing country, that of Latin America. Mexico City, as the largest city in Latin America, has some of the most severe problems.

Whatever the agglomeration economy benefits of large cities, the combination of unicentric spatial structure, environmental pollution and primacy render such cities undesirable. Urban primacy has been an especially marked characteristic of inward-oriented industrialization. Chapter 4 showed that urban primacy warps the evolution of the transport network, to the neglect of the remoter regions. By definition, rural areas as a whole must be more remote from high-level urban services under a primate spatial structure, than would be the case with a more hierarchical pattern in which more and smaller cities are spread through the country.

It is sometimes suggested that draconian measures are required to cope with the problems of emerging developing country megalopolises like Mexico City. These measures could include freezing the provision of infrastructure, as a 'cruel' solution which will force urban decentralization. A more cautious position would simply add in the full costs (including all environmental costs) which economic activity in very large cities incurs. This might be coupled with regional planning to promote decentralization. But the latter measure would be relatively ineffective, if macro and industrial policies were not reformed also, as Chapter 7 shows.

7

Restructuring the city and settlement hierarchy

Aspects of spatial restructuring

The growth of very large cities in developing countries has raised the question of whether spatial restructuring can ameliorate the resulting congestion. The economic historian E.A.J. Johnson (1970) has even gone so far as to argue that, 'It is my mature conviction that the problems of developing countries arise from the faulty organization and use of space.' Spatial restructuring may occur in either of two principal ways. First, it can occur at the local level by, for example, altering the internal structure of the city in order to improve its efficiency (Hall 1984). Alternatively, it can occur at the national level by modifying the settlement hierarchy, in order, for example, to reduce regional imbalances which favour a handful of very large cities (Friedmann 1973).

The internal structure of cities and the national size distribution of cities are obviously interrelated. If urban congestion can be successfully eased by adjusting the internal structure of the city, then the decentralization of urban activity to smaller cities loses some of its urgency. Or, vice versa, measures which increase the attraction of smaller towns and rural areas can retard rural–urban migration and thereby provide a breathing space for planners to tackle the backlog of congestion in the largest cities.

Proposals for spatial restructuring all too often under-estimate the forces working against them at the macroeconomic level, and such restructuring proposals have frequently been disappointing. More specifically, the policy of ISI encourages the location of economic activity close to the largest domestic market, which usually means the largest city. This strengthens the trend towards the dominance of a single city and consequently weakens the effectiveness of regional policies (Richardson and Richardson 1975). Meanwhile, at the local level, a highly unequal national income distribution skews urban investment towards those city districts (suburbs, business districts and recreational areas) which are favoured by the wealthy, as Walton (1978) shows with reference to Guadalajara. The poor may then take matters into their own

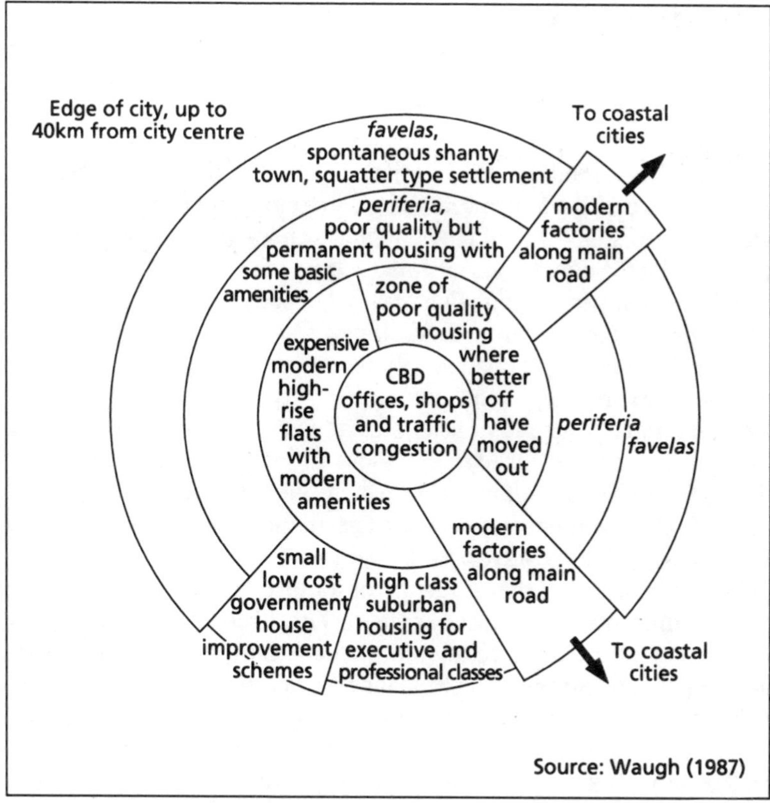

Fig. 7.1 *Internal structure of the Latin American city*

hands through, for example, the expansion of illegal squatter settlements. One implication of such spatial forces is that macroeconomic policy reform is likely to be a more effective means of alleviating urban problems than spatial restructuring. At the very least, the two sets of policies, macro and regional, should be synchronized.

This chapter examines spatial restructuring, drawing examples primarily from Latin America, the most highly urbanized major developing country region. It deals first with the internal structure of cities, outlining a model of urban structure and paying particular attention to the dominant land use, housing. The focus then switches to regional issues and the causes of the generally disappointing regional policy achievements, using Mexican and Venezuelan examples. Finally, the tension between macro and regional policies is brought to the fore, with reference to Brazil. An assessment is made of Brazil's efforts to alleviate the problems of inequitable land distribution and the dominance of São Paulo, through settlement of the Amazon.

City restructuring

Internal structure of the Latin American City

Figure 7.1 depicts a model of the internal structure of the Latin American city. A striking feature is the unicentric form. The city centre retains much of its importance, so that patterns of movement are characterized by long distances and cross-hauling. Such a unicentric structure may be appropriate for small towns and cities, but it becomes increasingly costly as city size rises. This is especially true in developing countries, whose cities leapfrogged the wheel-track era of the nineteenth-century city, shifting abruptly from the pre-industrial (or pedestrian) city to the automobile city. Consequently, most large developing country cities are deficient in rail networks which efficiently move large volumes of people through corridors of high-density settlement. They rely instead upon the automobile, which encourages low-density urban sprawl, so that movement involves a large number of vehicles over numerous and lengthy routes (Meyer *et al.* 1965).

An obvious solution to the problems of a unicentric city is the encouragement of a multicentric structure, as has occurred in many North American cities. A multicentric structure should increase the proportion of short local journeys. Fig. 7.2 illustrates proposals for restructuring Mexico City. The plan seeks to decentralize economic activity out of the old colonial core and into nine suburban centres at distances of 10 to 15 kilometres from the city centre. These nodes are to be linked by high density communication corridors along which run lines of the expanding subway system (Hall 1984). A buffer zone and conservation area is being developed to arrest further encroachment by the city onto the scenically attractive hills to the south (Ward 1990).

The infrastructure for a second set of nodes some 20 to 40 kilometres from the Mexican capital is being improved, notably with the provision of a high-speed motorway belt to the north and east. Finally, the adjacent state capitals such as Cuernavaca, Puebla, Tlaxcala, Pachuca and Toluca provide a third ring of nodes some 50 to 80 kilometres distant from the centre of Mexico City. Although such a multicentric system provides a coherent structure for reducing congestion within the capital city region, one obstacle to its realization is the fragmentation of responsibility between several state and local administrations. A further problem has been the sharp fall in the rate of investment, in the wake of the 1982 debt crisis (Lustig 1993).

Controlling housing provision: slums and public housing

Housing dominates urban land use, so that the control of its provision provides an important means of directing urban growth. Fig. 7.1 shows that

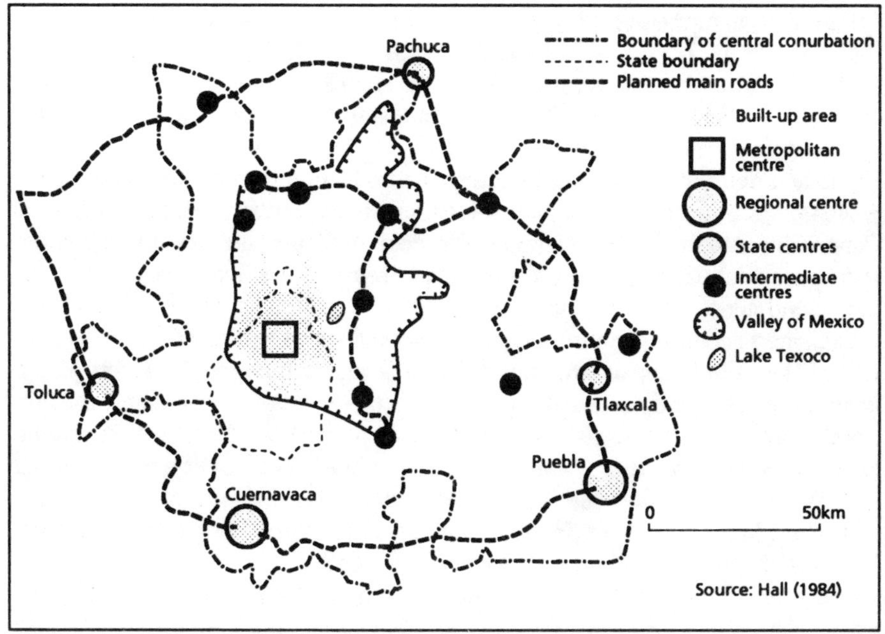

Fig. 7.2 *Restructuring of Mexico City*

in Latin American cities, upper class districts may abut the centre and extend in a sector towards the more attractive (downwind and elevated) suburban sites. The poorer housing is split between inner city slums, public housing districts, and shanty towns or *favellas*.

The inner city slums frequently comprise the former houses of the rich which have been subdivided (and allowed to deteriorate physically) in order to provide relatively cheap accommodation for the poor. The attraction of such accommodation for the poor is the access it gives to employment because of its proximity to the city centre and the public transport network. When formal urban jobs became more difficult to find during the debt crisis, and informal employment expanded by one-fifth to an estimated 30 per cent of the total (Gilbert 1994), inner city housing became much more desirable (Eckstein 1990).

But the quality of public services in slum districts is often poor, and hygiene and health may both be lower there than in the shanty settlements. A higher rate of communicable disease was detected, for example, in Kuala Lumpur slum settlements compared with squatter settlements (Aiken 1981), while a comparison of housing in Calcutta found that 57 per cent of multifamily groups in slums shared a single room, with each person having less than three square metres of space. There was one latrine and a single tap for every twenty people (Bose 1968).

Public housing is seldom provided on anything approaching a satisfactory scale; part of the reason is that standards have often been too elaborate. Many developing country planners set minimum standards of accommodation that are too high, and thereby price public housing out of the reach of those for whom it is primarily intended (Grimes 1976). Such housing space has typically ranged from 50 square metres in mid-income cities like Mexico City, down to 20 square metres in low-income cities like Nairobi and Madras. There has also been a preference for detached structures with individual services, which cost significantly more to provide than five-storey units with shared services.

Detached accommodation is typically beyond the reach of two-fifths to one-half of the population of low-income cities – just those people which public housing is designed to assist. In addition, the very poor may lack the job stability required to make monthly rental payments. As a result, public housing intended for the urban poor has often been taken by the middle class. Among the more successful developing country providers of housing for urban residents have been Hong Kong and Singapore. They had far smaller fractions of their population (less than 15 per cent) in slums or squatter settlements than most developing countries, which typically have 40 per cent or more. This difference arises because both Singapore and Hong Kong adopted low minimum standards for housing in the 1950s, and then steadily improved the quality of housing provision as per capita incomes rose and the resources became available to pay for it.

The provision of public housing is scarce not only because of over-elaborate design, but also because it yields a very low rate of return on capital and has a very long payback period. Typical ICORs for investment in housing are around 7, compared with 3 for investment in manufacturing. It has been argued that it is better to invest public funds in activities which yield the highest return, and to postpone housing investment until returns weaken elsewhere in the economy. Moreover, housing frequently has a substantial import content, estimated at one-third in the case of a typical African government housing scheme. A high import content strains the balance of payments.

But such unfavourable calculations ignore the wider social benefits of housing provision and the fact that public investment in housing may be highly advantageous, provided that housing specifications are pared back and local materials are used in construction. One attraction of house-building is that it is relatively labour-intensive, with a strong demand for unskilled and underemployed labour. There is a strong multiplier effect, with two indirect jobs being created by every direct job in construction (Grimes 1976). For example, the provision of house fittings stimulates small labour-intensive firms to supply them.

Additional social benefits emanate from the fact that houses provide not only homes, but also places of employment in many instances. For example, up to 30 per cent of the inhabitants of the Lima *barrios* are estimated to work at home. Residents may also secure income through renting rooms out. Additional social benefits arise from evidence that improved housing is associated with lower crime rates, better health and higher worker productivity. Such benefits are excluded from calculations of the financial return on public housing investment. But in order to ensure the maximum economic benefits from the provision of public housing, public housing needs to:

- draw mostly on local materials and suppliers;
- use simple techniques which can draw on unemployed labour;
- improve the spatial efficiency, by locating of new housing so that it is well-serviced and well-connected to places of employment.

In building housing, a trade-off is required between location, structure and services. The cost of land falls sharply towards the periphery of the city, while the height of structures means that, compared with a single-storey house, a unit in a five-storey block is 25 per cent more expensive, and one in a thirty-storey block is two-thirds more expensive. Overall, the cheapest public housing is a multi-family structure on the edge of the city with shared basic services (i.e. centrally located water and pit latrines and minimal security lighting). In Bogota, for example, multi-family structures with basic services would be affordable by 85 per cent of households on the edge of the city compared with 64 per cent at an intermediate location, and only 30 per cent near the city centre.

Controlling housing provision: squatter settlements

Because of the inferior quality of many slum dwellings and the inadequate provision of public housing, squatter settlements are frequently the solution for low income urban housing, by default. But their growth may weaken the control of planners over the urban structure. Such settlements typically house around two-fifths of the urban population and are thought to have originated in the 1940s. Contrary to appearances, they often reflect high levels of organization.

Leaders select the squatter families and then lawyers identify unoccupied idle land whose owners are unlikely to resist a squatter invasion. The positions of housing lots and public buildings are staked out at night, in advance of the movement of as many as 500 families, who rush to their allotted plot and erect crude dwellings. The squatters may subsequently invite friends and relations, so that infilling and expansion may occur. If the absentee landowner fails to contest the occupation, so that the squatters have security of occupation, such settlements are upgraded over a number of years. The initial structures are replaced by brick dwellings of more than one storey.

A study by Ward (1976) of squatter settlements in Mexico City made a distinction between the larger 'parachute' squatter settlements, of the type described in the previous paragraph, and the smaller ones, mainly to the immediate north and west of the city centre. The smaller settlements (*ciudades perdidas*) were typically rented accommodations with relatively high rental charges. The housing tended to be more cramped than in larger settlements and showed a low level of improvement. Their principal advantage, like the slum accommodation, appears to be their proximity to the city centre and the greater choice of employment.

Ward compared two larger squatter settlements, Sector Popular, founded in 1947, and Santo Domingo, established in the early 1970s, in order to capture the dynamics of such settlements. The population density of Sector Popular, at 750/hectare, was almost 3 times that of Santo Domingo because additional houses had been crammed onto the initial house plots. The older settlement also had many more two-storey structures, made of superior materials. Whereas the newer settlement provided a narrow range of basic services, such as a grocery, pharmacy and cafeteria, Sector Popular had a much wider range of activities. The overall picture is one in which the housing of the squatter settlement steadily improves in quality as incomes rise, a conclusion in line with the more optimistic view of urban poverty taken by Peattie and other critics of the marginalization thesis, discussed in Chapter 6.

An interesting attempt to combine the advantages of squatter settlements with city planning is provided by 'sites and services', a scheme pioneered, for example, in Venezuela's Ciudad Guayana in the 1960s. Elsewhere, in Manila, early experiments were not promising, but this appears to have been due more to the peripheral location of the scheme (which gave poor access to employment) and the tardy provision of basic services such as water and sewage disposal. More recent sites and services schemes in low-income countries such as Zambia and Tanzania have been more successful.

The basic idea of sites and services is that the government provides the land and divides it into plots which are levelled and given a concrete foundation with the room floorplan marked out. Basic services are provided by the state at modest prices, including water, drainage and power (which is often pirated from overhead cables by squatter settlers). Other services, such as schools, roads, clinics and waste disposal, are provided free of charge because of the strong external (social) benefits which they provide for the city as a whole. Small-scale industry is encouraged among the residences in order to diversify employment, a position which contrasts with that encountered by Eckstein (1977) in the Mexican suburbs, where such activity was illegal. In addition, care is taken to ensure that sites and services are adjacent to major job centres.

The residents of sites and services schemes are charged a nominal rent and given title to the land, so that they are assured of retaining the value of any house improvements which they make. Initially, the new residents make use of very crude building materials to construct simple structures, but as their incomes rise, recourse is made to brick and stone. Extra storeys can be added either for family accommodation or for rent to provide additional income. In this way, sites and services allow planners to regain control over the location of housing at relatively low investment cost, so that they can better direct the emerging urban spatial structure.

Restructuring the settlement hierarchy

Ward and Dubos (1972) proposed a three-step strategy to ease the problems of urban primacy in developing countries. The first step is rural land redistribution, along with more labour-intensive production methods. This is designed to retain more people on the land and thereby reduce the flow of migrants to the cities. In countries as diverse as Taiwan (Ho 1979) and Kenya (Tiffin *et al.* 1994) non-farm income has been important in raising rural living standards. Such activity has ranged from textile firms in rural Taiwan to tourist goods in rural Mexico (Kemper and Foster 1975).

The second step of the Ward and Dubos strategy is the stimulation of towns of intermediate size, which can act as 'schools' for migrants where they can adjust to urban lifestyles while being close to their families and traditions. Such migrants may commute back to their villages at weekends. The net effect of land reform and the expansion of small towns is to retard urban migration and (the third step in the strategy) to give the cities a breathing space in which to rectify the backlog of urban problems.

Land reform

Although land reform is considered to have been an important precondition for the successful industrialization of Korea and Taiwan, it has generally proved difficult to implement in most developing countries. Land reform has been especially problematic in Latin America, where the traditional communal ownership of land under the indigenous system (what Browning (1971) calls 'an inheritance of social equality') has given way to extremes of land concentration. For example, almost half the arable land in Brazil is owned by 1 per cent of all farmers, while in El Salvador in the 1970s some 263 families held 70 per cent of the land.

An unequal distribution of land is associated with rural poverty, especially in increasingly land-scarce Central America, where a shortage of land lies at the heart of socio-political conflict (Table 7.1). For example, in Tenancingo, El Salvador, the average size of holding among the younger generation of farmers was 1.1 hectares in the 1970s, compared with 1.4 hectares a generation earlier (Durham 1979). The monthly income on farms of less than 1.5 hectares was one-tenth that on farms of 7 hectares, while child mortality reached 48 per cent on the smallest farms, compared with 20 per cent on farms of 2.5 hectares or more. In Honduras, where the higher ratio of

Table 7.1 *Social and economic characteristics of six small Central American republics*

Country	Area (million sq. km)	Population (million)	Cropland/ hd (ha)	Population growth (%/yr)	Population in cities 1991 (%)	GDP growth 1980–91 (%/yr)	Per capita income ($1991)	Literacy (%)	Life expectancy (yrs)
Costa Rica	0.051	3.1	0.18	2.7	48	3.1	1850	93	76
El Salvador	0.021	5.3	0.14	1.4	45	1.0	1080	73	66
Guatemala	0.109	9.5	0.20	2.9	40	1.1	930	55	64
Honduras	0.112	5.3	0.35	3.3	45	2.7	580	73	65
Nicaragua	0.130	3.8	0.33	2.7	60	(1.9)	460	n.a.	66
Panama	0.077	2.5	0.24	2.1	54	0.5	2130	88	73

Source: World Bank (1993b), WRI (1992)

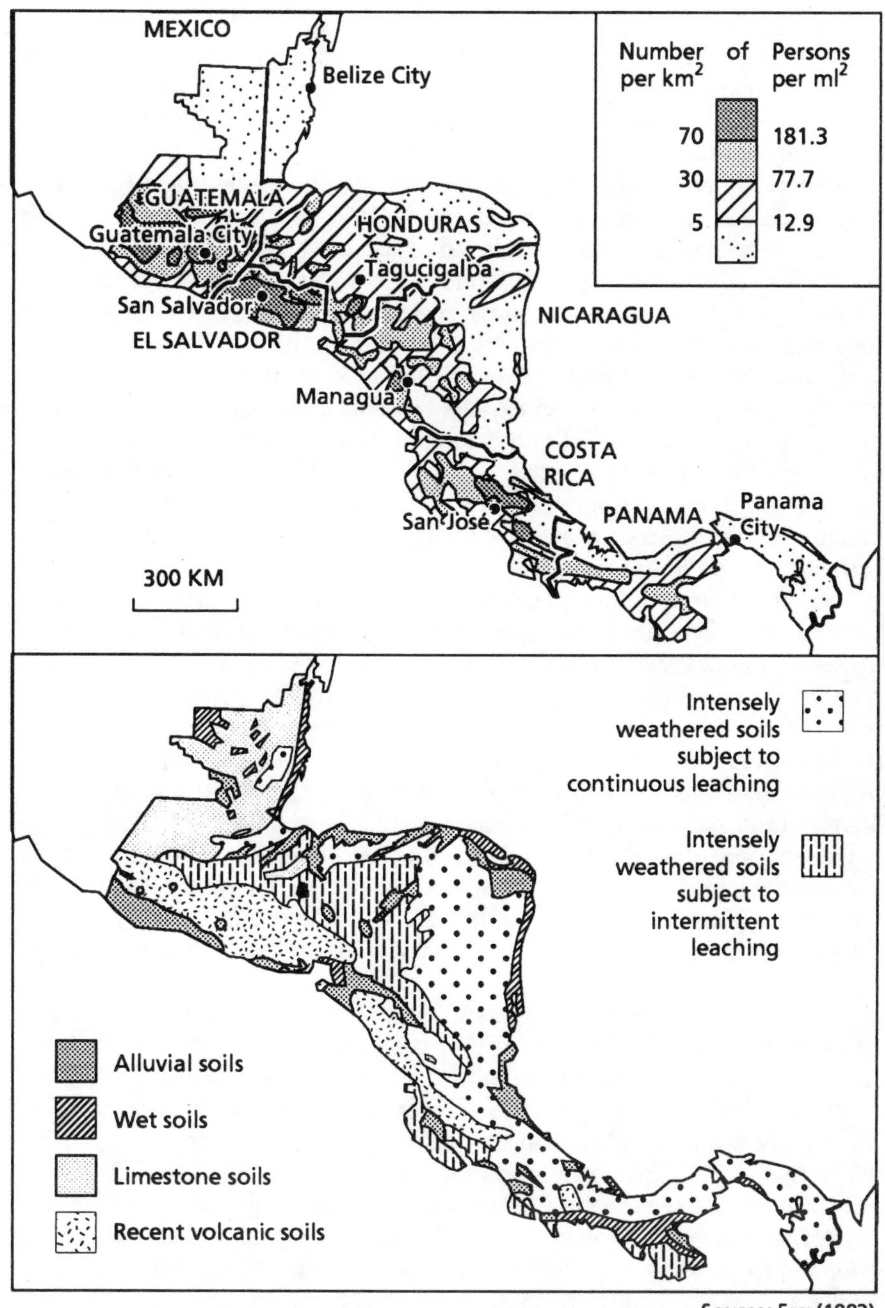

Source: Fox (1983)

Fig. 7.3 *Physical characteristics of Central America*

cropland per head is offset by poor soil quality (Fig. 7.3), the smallest farmers devoted 90 per cent of their land to growing food grains (mostly for subsistence), compared with only 6 per cent on larger farms. Yields were invariably lower on the smaller farms and declined at a faster rate, so that land required more frequent fallowing.

Yet agricultural success has not necessarily accompanied land reform. In Nicaragua, for example, the overthrow in 1979 of Samoza (whose family owned one-fifth of the arable land) led to around one-third of the country's land being redistributed as state farms and cooperatives. However, peasants disliked working on the state farms because, as in Maoist China, they felt they were not adequately rewarded in relation to the effort they expended. The newly elected government of El Salvador in 1980 set a 250-hectare maximum limit on farm size, and granted tenants and share-croppers the right to buy up to 7 hectares. Yet only one-third of the 150 000 farmers expected to benefit from this policy did so, due to problems of implementation. Similarly, in Brazil, a programme for the compulsory purchase and redistribution of under-utilized land to 1.4 million families had achieved only 5 per cent of its target in the late-1980s.

Even in the case of the more ambitious Latin American land reforms such as those of Bolivia, Peru and Mexico, the inadequate provision of credit to the new owners created a peasant farm sector which lacked the dynamism required to prevent a relapse into abject poverty, as population pressure persisted. In Mexico prior to the 1910 revolution, an estimated 97 per cent of the land was owned by 1 per cent of the farmers and, although millions of hectares were subsequently redistributed as *ejidos* (small plots, to be individually worked but communally owned), the 6.5 hectare average size proved too small. Consequently, the diffusion of green revolution techniques during 1945–65 benefited the large farms, but production on smaller units grew much more slowly, due to lack of irrigation and credit. By the 1980s, some three-quarters of the country's poor, accounting for one-fifth of the population, lived in rural areas.

The prospects for Mexican peasant farmers did not improve. Plans made during the 1979–81 oil boom to trigger a second green revolution for small farms through the provision of subsidized fertilizer fell victim to the debt crisis. Subsequently, the 1993 entry of Mexico into the North American Free Trade Agreement signalled a further retreat for small farmers. The phasing out of the 70 per cent import tariff on maize, coupled with the right to sell *ejido* land, will hasten the consolidation of farms into larger, more viable units. But for such a policy to work, Mexico's shift to outward-oriented industrialization must generate sufficient urban jobs to absorb those leaving the land. Similarly, the six small republics south of Mexico will gain more from a revival of the Central American Common Market (which rendered manufacturing sufficiently dynamic, before the collapse of the agreement in 1969 to draw surplus labour off the land) than from either land reform or farm subsidies.

Intermediate settlements

Given the implementation difficulties of the land reform component of the Ward and Dubos strategy, Johnson (1970) has prudently concentrated on the role of towns in restructuring of the settlement hierarchy in developing countries. He identified three basic types of settlement pattern, which are:

- primate (the dominance of one or a handful of very large cities);
- socialist (with fairly self-sufficient and self-directed communities of 20 000 to 50 000, with basic control from the political centre);
- hierarchic (the Christalleran central place hierarchy of nesting large, medium and small service centres).

Johnson favoured the hierarchic system, because it was the most effective of the three in bringing modernization to all parts of a country, including the remoter countryside. Citing the historic success of the USA, Denmark and Japan, Johnson argued that a hierarchic settlement structure maximizes the division of labour, regional specialization and trade.

Johnson rejected the primate spatial structure not only because he felt that the diseconomies of congestion far outweighed the agglomeration economies, but also because the village is an inefficient catalyst for change. Drawing on Indian examples, Johnson argued that the village lacks sufficient size to support a bank, so that lending is monopolized by the local elite, who not infrequently collude to charge annual interest rates of several hundred per cent. Similarly, the village provides an inadequate market for the harvested crops because there are too few buyers, so that, once more, the wealthier farmers collude to set low prices. Villages are also too small to generate the traffic to support either an all-weather road (as Owen (1968) notes), or the range of talents required to diversify the rural economy into industry. A further factor here is the strong social pressure within small communities to conform to established behavioural norms which, according to Johnson, curbed personal initiative.

Rather, the maximization of economic opportunities required the development of towns, the deficient component in the primate structure and the element required to transform it into a hierarchic pattern. For example, whereas the ratio of towns to villages in Europe was 1 to 16, that in Egypt was 1 to 100 and in India 1 to 185. Johnson rejected a communal organization of space, on the grounds that moral incentives fail to motivate individuals and centralized bureaucracy stifles individual initiative and creativity. He cited as evidence the failure of socialized systems in the Soviet Union and Eastern Europe.

Johnson's views grew out of his strong right-wing conviction that pioneering nineteenth-century America offers a model for economic progress. Countries as diverse as Israel and Kenya have acted on his thesis, which is, however, particularly appropriate for low-income countries in the early stages of development where the rural sector needs to be brought into the commercial economy. But the mid-income developing countries like those of Latin

America have in many cases already developed a strongly primate spatial structure. The problem for them is therefore less one of avoiding regional imbalances, than of correcting those which already exist.

Regional policy in mid-income developing countries

The cause of severe regional imbalances

Williamson (1965) argued that as per capita incomes rise, so regional income differentials initially widen and then narrow. The widening regional income gap reflects the operation of agglomeration economies in the core region and the backwash effects on the periphery (discussed in Chapter 6 with reference to migration). Using the coefficient of variation to yield a standardized measure of the variation in regional incomes about the mean, Williamson showed that whereas a low-income country like India had a low index (0.17) that was also within the range exhibited by the industrial countries, the mid-income countries of Latin America had much higher indices. The indices for Brazil, Mexico, Argentina and Venezuela were 0.6 or more, but that for Colombia (where Cali and Medellín rival Bogotá and curb its primacy) was 0.24.

According to Williamson, the regional income differential will eventually narrow as congestion diseconomies in the core depress the efficiency of capital there and enhance the attractions of the erstwhile peripheral locations. State intervention to minimize income differences may be an additional factor closing the regional income gap. But this optimistic view can be challenged on the grounds that regional income amplification on the scale observed in much of Latin America is not an inevitable outcome of development, as countries like South Korea show. Rather, such high inequality reflects the operation of strong political forces, which first create excessive income divergence and then delay convergence (Gilbert and Gugler 1992).

The adoption of a policy to reduce regional income differentials can be justified on efficiency grounds as well as for reasons of greater social justice. The efficiency gains arise from the reduction of the diseconomies of congestion in the core and a more efficient use of capital and labour within the lagging region (Armstrong 1993). But it is debatable whether initiatives pitched at the regional level will be effective in the absence of fundamental changes in macroeconomic policy. For example, Storper (1991) has argued with reference to Brazil that polarization results from a combination of sluggish economic growth and unevenly skewed income distribution. In contrast, he considers that decentralization requires rapid economic growth based on widening purchasing power. In addition, Storper maintains that decentralization requires the emergence of a new 'ensemble' of industries (high-tech, in the case of Brazil). The new ensemble draws its competitiveness from the localization economies (i.e. industrial product-specific scale economies) which take root away from the more generalized agglomeration economies that anchor earlier ensembles, such as petrochemicals or engineering, to the established major cities.

It is refreshing to have regional policy discussed in the context of macroeconomic policy, but Storper's case is weakened by the fact that his structuralist stance dismisses orthodox views, and so neglects the East Asian development model. That model suggests that both class and regional inequality can be minimized by avoiding the inward-oriented industrial policy favoured by most Latin American countries. But such lessons come too late for Latin American countries, many of which now have sizeable regional income gaps as a result of the flawed policies they have pursued. Nevertheless, Latin America has considerable experience with regional policy instruments such as grants, worker rebates, tax incentives and growth poles. Their success can be judged with reference to Mexico, Venezuela and Brazil.

Half-hearted regional policies in Mexico

Mexico has deployed several regional policy instruments, in an attempt to offset the concentration of industrial activity on Mexico City and the northern city of Monterrey. In 1937 state exemptions were introduced on land taxes, stamp duty and sales taxes, but because all states eventually abolished them, the areal benefit was neutralized and proved ineffective. In fact Mexico City and Monterrey restored the taxes in 1954, because the exemptions had reduced state revenues by 3 per cent and most firms could recoup the cost of the taxes with a modest 1 per cent rise in prices (Lavell 1972).

Other regional incentives, such as the 1941 New and Necessary Industry Law, and the Fund for Small and Medium Factories, proved similarly ineffective, as did the river basin schemes sponsored in the remote north west of the country. In the case of the river basin schemes, Barkin (1975) reports that the Tepalcatepec Basin scheme provided local small farmers with dams, irrigation and roads to encourage a switch into cotton production for export. But the locals lacked the capital to begin production and so rented their plots to wealthy outsiders, for whom they then worked as hired labour. When the price of cotton declined in the 1960s the outsiders left, having done little to diversify the local economy in the mean time.

After the 1960s, Mexico turned to growth poles to spur regional development. The growth pole theory is usually traced back to Perroux and assumes that a key industry which is located in a lagging region will trigger backward and forward linkages and stimulate growth throughout the region. Von Boventer (1975) has argued that growth poles can exploit two possibilities: either proximity to an established core, or extreme remoteness from such a core. In the first case, the growth pole captures the agglomeration economies and reflects the spatial restructuring being attempted, for example, around Mexico City (Fig. 7.2). In contrast, the remote pole exploits the spatial monopoly which is conferred on it by virtue of high transport costs from the core. Alonso (1971) stresses the importance of the size of the growth pole: remote poles need to be large, to establish the scale economies, whereas poles forming part of a multicentric core may be smaller.

Mexico shows that an important cause of growth pole failure is the use of too many small poles, which squander the scale effects (Scott 1982). In the 1970s Mexico established a four-tier hierarchy of growth poles, with assistance being given to the lowest two tiers. The unassisted tiers comprised Mexico City at the peak, and Monterrey and Guadalajara in the second level. Thirteen additional regional metropolitan centres were designated as growth poles, including five in the Central Valley, two in the north and six along the Gulf Coast. With the fourth tier, this created too many small poles.

Richardson and Richardson (1975) and Friedmann (1973) agree that large poles appear to be essential for success, and in the 1980s Mexico adopted four large growth poles: two on the Gulf Coast and two on the Pacific Coast. These large new poles are much more likely to be effective than the earlier small ones. This is not only because they will be large, but also because their coastal location will benefit from Mexico's shift towards an outward-oriented, export-led development strategy since the mid-1980s.

Venezuela's resource frontier growth pole

Earlier Venezuelan experience with large growth poles, however, cautions that they can be difficult to implement and that they may yield a very poor return. In the 1950s, the Venezuelan capital city and the surrounding highland area dominated a largely empty country. To the north west lay the arid coastal zone, and the savannas and forests stretched to the south. The restoration of democracy in 1958 triggered plans to restructure the space economy.

The Venezuelan government called upon a team of planners from MIT and Harvard University to devise a suitable regional plan. Influenced by Friedmann, the planners opted to establish two industrial cores, to counter the dominance of Caracas. One core comprised a petrochemical complex straddling the neck of the Gulf of Maracaibo, while the second core was a metallurgical complex in the eastern savannas at the junction of the Caroni and Orinoco Rivers (Fig. 7.4).

The two large cores reflected the conviction of Friedmann (1966) that large size is necessary to capture agglomeration economies and to maximize the opportunity for human interaction and, therefore, innovation. In addition, a large core would secure sufficient political power for the new core to successfully challenge Caracas for financial resources. It was hoped that by carefully locating the cores in lagging regions, the extent of the 'upward transitional areas' (which benefit from the trickle-down effects of the core)

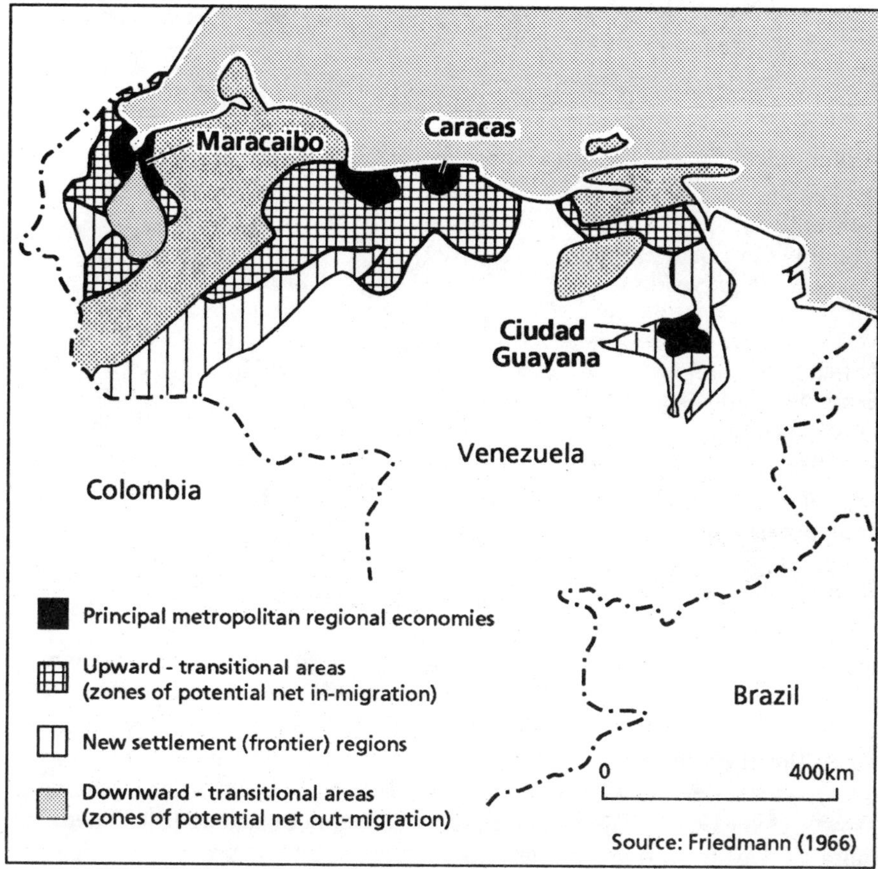

Fig. 7.4 *Venezuelan growth poles*

would be maximized and the 'downward transitional areas' (where backwash effects would be strongest) would be minimized.

The Ciudad Guayana core was targeted to have a population of 350 000 by 1980, producing 4.5 million tonnes of steel, 21 000 tonnes of aluminium and 2000 MW of hydroelectricity. It would create 100 000 new jobs (Table 7.2), of which one-third would be in heavy industry, one-quarter in light manufacturing, farming and construction, and the remainder in services. Heavy industry would receive two-thirds of the investment and generate a 15 per cent return, which would be used to expand the industrial complex and to trigger new growth poles in frontier regions elsewhere in the country. A state corporation, CVG, was set up to implement the project, with administrative powers over a large geographical area.

The scale of the Ciudad Guayana growth pole was sharply increased in the 1970s, so that its implementation took the form of an HCI big push, with all

Table 7.2 *Guayana region employment structure, 1965 projection and 1984 outturn*

Sector	Projection (1965)	Outturn (1984)	Actual/ projection
Heavy industry			
Mining	6900	4542	0.66
Energy (and utilities, 1984)	900	4171	4.63
Heavy manufacturing	26 000	22 658	0.87
Total	33 800	31 371	0.93
Agriculture	5500	5010	0.91
Secondary industry	14 ,000	5144	0.37
Construction	7600	17 893	2.35
Services	29 100	117 957	4.05
Total employment	100 000	177 375	1.77
Total population	357 000	753 172	2.11

Source: Auty (1990a)

the risks which that implies (see Chapters 5 and 6). The hydro scheme was quadrupled to 8000 MW and aluminium capacity was increased to 400 000 tonnes and integrated backwards into an alumina refinery and later a bauxite mine. But the growth pole proved over-ambitious and was plagued by cost over-runs and severe start-up difficulties, which depressed national economic growth (Auty 1990a). The steel plant cost $5 billion to construct – 50 per cent more than planned – and experienced severe technical problems which cut operating capacity and caused heavy losses. Domestic demand proved insufficient, so steel was exported, despite its high costs. Although the aluminium investments gave a better return, this owed much to the supply of very cheap hydro-electricity, which did not earn an adequate return on the $7 billion investment.

Consequently, far from generating the expected 15 per cent return, the Ciudad Guayana metals complex acted as a drain on the country's investment resources through much of the 1980s. Although the city's population reached 75000 by the mid-1980s – twice the target figure – only one-quarter of the residents lived in the new city; the rest occupied squatter settlements on the east bank of the Caroni River. Moreover, most jobs were in construction (due to delays in project completion) and in services, because the heavy industry failed to stimulate the downstream light manufacturing expected (Table 7.2).

Moscovitch (1969) calculated that if the investment in Ciudad Guayana had been applied to farming in the Orinoco Delta, it would have generated five times as many jobs as the heavy industry complex and one third more output. Although the return on investment would have been lower (11 per cent), the project would have been less risky and easier to scale back than the HCI growth pole, whose two largest projects, the hydro dam and steel plant, cost $12 billion. It is difficult not to conclude that the large Venezuelan growth pole was over-ambitious and that it represents a massive misallocation of capital investment.

The three Brazils

Brazil's regional policy echoes those of Mexico and Venezuela: it was designed to alleviate development strategy failings, and proved largely ineffective. The settlement of Brazil by European colonizers has been characterized by the dependency theorist, Gunder Frank, in terms of a series of booms, beginning with the sixteenth-century sugar boom in north-east Brazil. That was followed in the eighteenth century by a mineral boom in the state of Minas Gerais, which in turn gave way in the nineteenth-century to the coffee boom around São Paulo. The linkages from the coffee boom were relatively favourable for local economic development (Hirschman 1977), and São Paulo became the hub of a major industrial expansion which swelled the population of the city to 20 million and left the state of São Paulo with more than half of Brazil's manufacturing output (Fig. 7.5).

A corollary of the rise of south-eastern Brazil is the relative decline of the north east. The two eastern regions maintained rough parity in population until the late nineteenth century, each having around 45 per cent of the country's inhabitants. Thereafter, however, the south east became dominant (Baer 1989). By the 1980s the south east, which occupies 16 per cent of the Brazilian land area, accounted for 58 per cent of the population, 76 per cent of the GNP and 92 per cent of manufacturing. Meanwhile, the north east has 18 per cent of the land area, 29 per cent of the population but only 14 per cent of GNP and 7 per cent of manufacturing. The remaining sparsely settled north-west frontier region occupies two-thirds of the country but boasts only 13 per cent of the population and 1 per cent of manufacturing output (Fig. 7.5).

By the 1950s, the gap between the industrializing south east and the poverty-stricken north east was causing massive out-migration, which involved 10 per cent of the population of the north-east in that decade. The out-migration was swollen by periodic droughts in the arid Sertão region which lies inland from the plantations on the humid coast. Such droughts typically elicited initial concern from the federal government, which provided one or two years of emergency investment for infrastructure; this aid was then cut back, leaving unfinished the dams and irrigation schemes that were intended to alleviate future droughts. The severity of the 1958 drought led to the formation of a regional agency, SUDENE, with responsibility for restructuring the rural and industrial economy. However, apart from the expansion of some manufacturing activity in Salvador and Recife, progress was minimal and out-migration continued.

	North East	Brazil total
Population	42m	146m
Life expectancy	46	60
% households below poverty line	44.0	23.3
% households with electricity	34.3	60.2
Infant mortality	100 in 1000	60 in 1000

	Population (m)	% of total	% of total GDP
SOUTH EAST	**62.2**	**42.5**	**58.2**
São Paulo	32.0	21.4	36.0
Rio	12.5	8.6	12.0
SOUTH	**22.1**	**15.1**	**17.7**
NORTH EAST	**42.2**	**29.0**	**13.6**
NORTH WEST	**19.5**	**13.0**	**10.5**

South/south east total: % of total population; 58 GDP; $200bn
North/north west total: % of total population; 42 GDP; $63.5bn

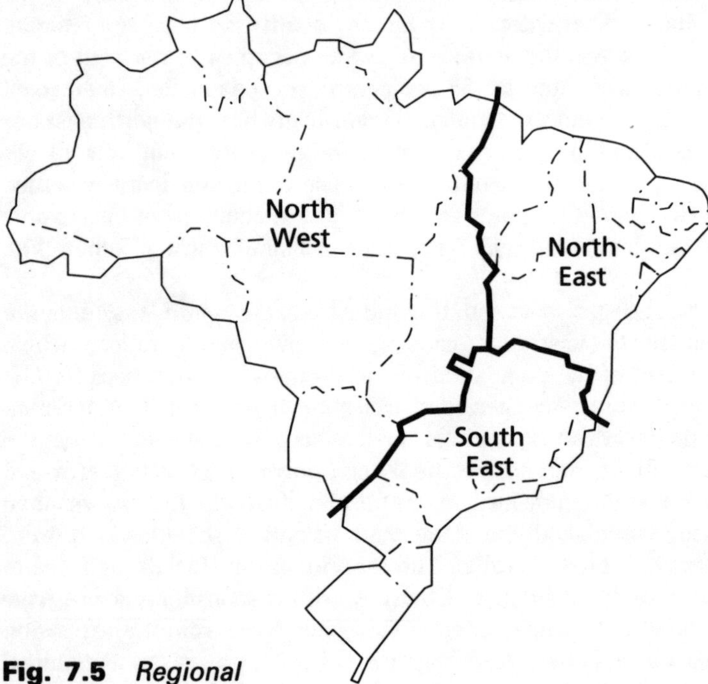

Fig. 7.5 *Regional contrasts in Brazil*

Source: *Financial Times* (1992)

Attention centred increasingly on the north-west region as an alternative destination for migrants from the poverty-stricken north east. An ambitious programme of road-building commenced which opened up large tracts of the Amazon and the adjacent savannah regions to peasant farmers, ranchers and mining companies. In addition to the construction of Brasilia on the Mato Grosso savanna in the late-1950s, fifteen agro-mineral complexes were designated as growth poles in the Amazon in the 1970s (Kleinpenning 1977). The most important of these poles is the projected $60 billion Gran Carajas iron-ore mine and hydroelectric plant. At the same time Brazil also built a large petrochemical plant at Camacari, some 25 kilometres north of Salvador.

Some success was achieved: economic growth in the north east exceeded that of the south-east from the mid-1970s, and the per capita income of the north east rose from 40 per cent of the national average to 55 per cent. But capital-intensive growth poles like the Camacari petrochemical complex and Carajas have weak regional linkages and contribute little to manufacturing employment. For example, the construction of the Camacari plant required federal subsidies and capital from the south east. It imports most of its raw materials from the south east, where most of its production is processed (World Bank 1982).

The growth poles in the north-west region are also controversial and of limited effectiveness. Brasilia has moved several million consumers closer to the rice and beef farms of the Mato Grosso, but created little manufacturing, while the region has been exposed to competition from São Paulo industry (Katzman 1977). Meanwhile, the exploitation of the Amazon region has elicited increasing concern over the environmental damage (Eden 1978, Mather 1989) as well as for the welfare of native Indians and peasant settlers (Smith 1981). Moreover, many large investments, including Carajas, have proved over-ambitious because of slower growth in Brazilian and world demand, and poor performance (de Sa and Marques 1985).

The rash exploitation of north-east Brazil reflects the failure of both industrial policies and land reform in the eastern part of the country. First, the autarkic industrial policy which Brazil adopted strongly benefited the south east. As already noted, most manufacturing is located there, at the expense of the north east (Goldsmith and Wilson 1991). The comparative advantage of the north east in cheap labour would have attracted industrial investment, had not Brazil leap-frogged the second stage of the East Asian development model and favoured ISI. Meanwhile, slow-maturing HCI earned insufficient foreign exchange, so that the need to secure additional export revenues was a key reason for Amazonian investment in the 1970s.

Second, pressure for land redistribution in the settled eastern part of the country was reduced by the availability of vast areas of land in the Amazon. The Brazilian government encouraged the in-migration of small farmers, who have been the major agent of deforestation (Pearce *et al.* 1990). But the government also adopted fiscal measures which encouraged ranching –

the second most important cause of deforestation in the Amazon – even though the revenues earned by ranches cover less than half their costs. The tax loopholes made the ranches worthwhile because they had the effect of transforming what otherwise would have been socially unprofitable projects, into privately profitable ventures (Binswanger 1991). Such tax incentives are also massively regressive, because they favour the wealthiest in Brazilian society.

It is clear that with a sound national development strategy, Brazil would not have needed to exploit the Amazon, and that even now, policy reform would go a considerable way towards relieving pressure on the Amazon region. Brazil provides yet another Latin American case where regional policies have been pursued in order to mask the consequences of flawed macro policies, and have failed to work. It also shows how a bountiful resource endowment can be used to postpone desirable policy reform, and how that only serves to make the latter objective even harder to attain. Brazil in the 1990s is a classic illustration of the resource curse thesis.

Conclusion

Large regional variations in income are not a necessary product of economic growth, as Korea shows. Rather, they reflect policy errors. In particular, autarkic industrial development concentrates wealth on upper income groups and on a few major cities. Consequently, urban and regional policies which seek to counter the adverse effects are unlikely to be effective, as long as the flawed macro policies persist.

The spatial restructuring which the use of large growth poles involves has not been particularly successful in Latin America, or elsewhere (Auty 1990a; Rondinelli 1991). The growth poles have tended to be capital-intensive, with minimal local linkages and their economic contribution seems likely to take more than one generation, rather than a decade, to bear fruit. But growth poles also demand large investments for their infrastructure needs, and the scale of individual projects is not easily scaled down if markets unexpectedly deteriorate. Both Venezuela and Brazil show that such investments may well give poor returns, on a scale that can adversely affect national economic growth. In addition, their capital-intensive nature contributes little to useful job creation in the lagging region.

In the case of Brazil and Mexico, it is difficult to escape the conclusion that the exploitation of the Amazon and oil, respectively, was not required in order to raise living standards. Rather, both countries (and others within the region) used their rich resource base in order to avoid taking hard political decisions about redistributing land to help small farmers, and about the reform of industrial policy.

Meanwhile, Korea, a country with one-quarter of Brazil's population (implying a much smaller domestic market for industry) and one-hundredth of its land area (and therefore far fewer natural resources) has industrialized much more rapidly and with far less income equality. The Latin American experience is an interesting example of the resource curse thesis, which is explored in more detail with reference to the mineral economies in Chapters 8 and 9.

PART 5

Hard lessons from the new international economic order: mineral economies

8

Critiques of international trade

All three groups of developing countries examined thus far have tended to favour inward-oriented policies, and trends in their economic growth and income distribution have been disappointing. This chapter seeks to explain why so many developing countries were suspicious of the prevailing liberal international economic order (LIEO). It reviews changing postwar attitudes towards trade and aid, and analyses the role of state enterprises and producer country cartels as instruments for changing the LIEO.

A prime goal of the LIEO was to open up domestic markets to free trade. It was assumed that this would allow all countries to specialize in production in line with their comparative advantage, and thereby improve the efficiency with which they used their capital, land and labour (Williamson and Milner 1991). The apparent success of postwar moves towards freer trade caused some orthodox economists to refer to the years between 1950 and 1973 as the 'golden age' of economic growth. During these years, economic growth rates exceeded those of the earlier high-growth period of 1870 to 1913, as barriers to trade between the industrial countries shrank (Lewis 1978). But, as noted in Chapter 1, this optimistic view of postwar development is disputed by the structuralists.

The structuralists criticize what they see as the adverse impact of trade on the developing countries, and the unsatisfactory nature of the aid provided by the industrial countries. The critics felt that the industrial countries derived substantial benefits from trade, at the expense of the developing countries, and that the aid which they donated to developing countries provided inadequate compensation. In the late 1960s and early 1970s the multinational corporation (MNC) was singled out for particular attack. The MNC was viewed by some observers as the latest in a line of institutions, stretching back to colonialism, by which the rich countries controlled the developing countries.

Discontent with the LIEO had its roots in work undertaken in the 1950s by Singer (1950) and Prebisch (1950). It may be recalled from Chapter 1, that their views are associated with the diffusionist wing of the structuralist group and call for reform of the LIEO in favour of the developing countries. A more radical structuralist approach emerged in the 1960s in the form of dependency theory. This concluded that an improvement in the lot of the developing countries required radical change of the global economic system.

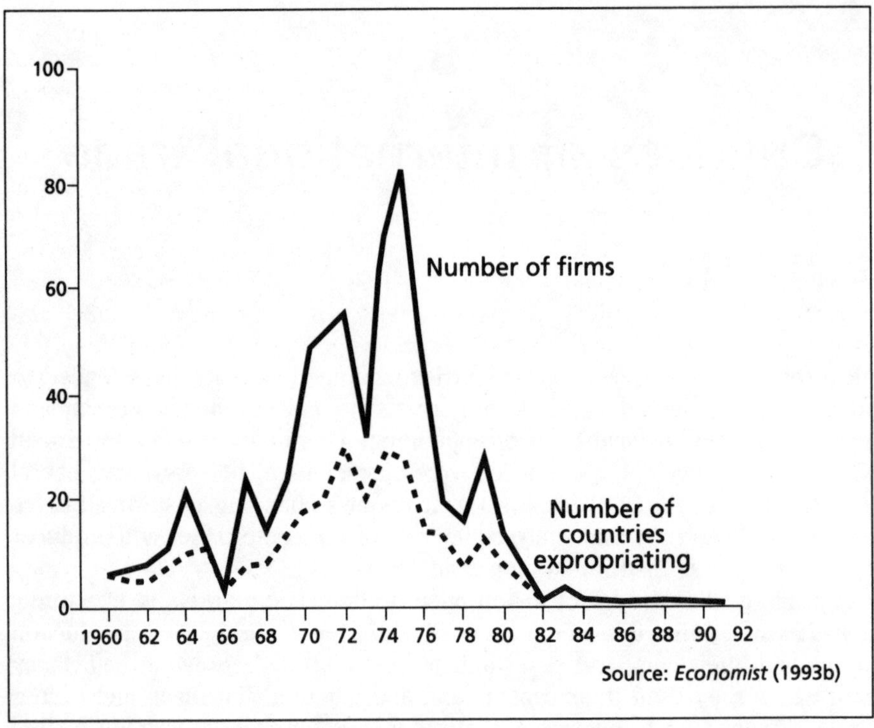

Fig. 8.1 *Expropriation of MNC assets, 1960–90*

The criticisms of both the reformist and radical structuralist viewpoints underpinned a growing demand for a new international economic order (NIEO) in the 1960s and 1970s. Two important objectives of the NIEO were the formation or strengthening of producer country cartels, such as OPEC, and a reduction in the power of MNCs. Producer cartels were championed in order to achieve higher prices for the primary product exports of the developing countries. As for the MNCs, nationalization of their developing country subsidiaries, which had begun in the 1950s (Williams 1975), accelerated into the 1970s (Fig. 8.1).

Mining MNCs were especially singled out as targets for nationalization, because they often functioned as economic enclaves. This meant that MNC mines had minimal local linkages and gave maximum economic stimulus to distant regions. Mines provided the clearest example of a once-for-all depletion of natural resources, which opponents of the LIEO claimed occurred with insufficient economic benefit to the host developing country.

The formation of producer country cartels and the nationalization of MNC subsidiaries were both manifestations of the postwar rise in the level of state intervention in economic management within the developing countries, which also included the adoption of varying degrees of industrial autarky. The rise in state intervention was accompanied by a sharp increase in the level of direct lending by western banks to the governments of the developing countries. That lending was boosted after 1973, as the banks sought to recycle the petrodollars deposited with them by the newly rich oil-exporting countries. The surge in lending to developing country governments was at the root of the subsequent 1982 debt crisis, discussed in Chapter 10.

The increased capital inflow highlighted a new set of problems for the developing countries, because by the early 1980s it was clear that the increase in capital resources would not resolve their problems. The new challenge was how to effectively deploy the extra resources for national economic development as opposed to the short-term political gain of minority groups within the developing countries. The mineral economy group of countries provides a particularly clear illustration of this problem, and its experience is discussed in detail in Chapter 9.

The present chapter explains why discontent with the LIEO led to increased economic intervention by most governments in the developing countries. It begins with a review of the Singer–Prebisch thesis concerning the unequal participation of industrial and developing countries in international trade. It then focuses on the more radical dependency perspective with particular reference to the case against the MNCs. The second half of the chapter then examines some of the consequences of the state intervention which the structuralists advocated. It does so with reference first to the role of producer country cartels, and then to the performance of the state-owned enterprises (SOEs) which replaced many of the MNC subsidiaries.

The Singer–Prebisch critique of international trade

Unequal exchange: trade and aid

During the 1960s both the developed and developing countries were growing rapidly, with annual economic growth rates in the latter typically at 6 per cent, compared with 4–5 per cent in the industrial countries. But high population growth rates in the developing countries meant that their per capita income grew more slowly than that of the industrial countries. For example, per capita income grew in the developed countries in the 1960s at 4.1 per cent per annum, compared with only 1.8 per cent for the low-income countries of sub-Saharan Africa and South Asia.

Such trends gave rise to projections that the per capita income of the OECD countries would double to $10 000 (in constant 1975 dollars) between 1975 and the year 2000, whereas that of the poorest developing countries would rise from a meagre $70 to only $100. The income gap between rich and poor seemed to be inexorably widening. Already in 1980 the industrial

Table 8.1 *Official aid 1992 (%GDP)*

Country	Aid	Country	Aid
Norway	1.2	Belgium	0.4
Denmark	1.0	Australia	0.4
Sweden	1.0	Italy	0.3
Holland	0.9	Britain	0.3
France	0.7	Japan	0.3
Canada	0.5	Spain	0.3
Switzerland	0.5	USA	0.2
Germany	0.4		

Source: OECD

countries accounted for 77 per cent of GWP (gross world product) but only 27 per cent of the global population, whereas the low-income countries of Asia and sub-Saharan Africa comprised 49 per cent of the population but only 5 per cent of GWP.

International trade provided the structuralists with a plausible explanation for the widening income gap. The developing countries relied heavily on the industrial countries for trade: some 75 per cent of the trade of the developing countries was with the industrial countries; only 25 per cent was among themselves. The exports of the developed countries were mainly manufactures before the 1970s, whereas primary products like minerals and tropical crops dominated the exports of the developing countries (manufactured goods accounted for barely one-fifth of their exports). During much of the postwar period, the prices of primary products fell relative to the prices of manufactured goods. Development economists like Myrdal (1957) joined Singer and Prebisch to argue that the global trading system favoured the developed countries at the expense of the developing countries. The developing countries were urged to raise protective barriers against the import of manufactured goods from the industrial countries, in order to build up their own industrial capabilities.

A second major criticism of the LIEO concerned the level and nature of the aid transferred from the rich countries to the poor countries. Not only were the industrial countries seen to reap most of the benefits from trade, they were also becoming increasingly mean with their aid as they got wealthier. In the 1960s, the UN set the OECD countries the target of raising the annual level of their aid to 0.7 per cent of GNP. Yet all but the Scandinavian countries retreated from the target, between 1960 and 1971. For example, French aid fell during the 1960s from 1.38 per cent to 0.67 per cent of GNP, while Belgian aid fell from 0.88 per cent to 0.5 per cent, and that of the UK, the US and Japan dropped to less than half the target level (Dinwiddy 1973). By 1975, OECD aid averaged only 0.35 per cent of GNP, and although aid increased by 60 per cent in real terms through the 1970s, it continued to decline as a fraction of GNP (Table 8.1). Yet influential intellectuals, such as Lord Snow and the Russian physicist, Sakharov, pleaded for a rate of aid donation of 20 per cent of GNP annually.

Some three-quarters of all aid came from the OECD by 1980, compared with 18 per cent from OPEC and 6 per cent from the centrally planned economies (which denied any responsibility for the relative impoverishment of developing countries). Aid was less than one-tenth the value of OECD exports to the developing countries. It was regarded as unsatisfactory not only because it was low, but also because it was tied in a way which reflected the strategic interest of the donor. For example, in the 1960s, low-income India received $2/head compared with $45 for Malta, a more prosperous but strategically located country. More than two-thirds of the aid was tied to spending in donor countries, often on capital-intensive equipment rather than locally produced items, or imports of the recipient country's choice. Meanwhile, OPEC aid went overwhelmingly to Muslim countries. For these reasons (with the honourable exception of the Scandinavian countries), as the rich countries became richer, they were easily portrayed as benefiting from trade while becoming ever meaner and more manipulative with aid.

The diffusionist response

As disillusion set in during the 1950s with the failure to achieve the initial optimistic postwar goals in Latin America (see Chapter 6) and later in sub-Saharan Africa, the LIEO increasingly became the target of attack. Prebisch argued that trade was retarding the diffusion of economic benefits to the developing countries. The world economy comprised an industrial core and a developing periphery. The latter specialized in the export to the core of primary products such as minerals, and tropical crops like coffee and sugar cane. Unrestrained global competition resulted in a disproportionate rate of accumulation of capital (surplus) in the industrialized core countries.

One reason for the gains made by the rich countries was the low elasticity of demand (that is, the rate of growth in demand for such products relative to the rate of growth in purchasing power) for most tropical products. Demand for primary products grew during the postwar years at only 85 per cent of the rate of increase in the demand for traded manufactured goods (Lewis 1978). A second cause of the declining prices of primary products relative to the prices of manufactured goods was the simultaneous entry of many new tropical producers into world trade during the postwar years. This created intense rivalry in selling similar products to a limited industrial country market, which depressed prices. For example, new East African tea and coffee growers captured markets at the expense of established producers in South Asia and Latin America.

Prebisch argued that the industrial economies controlled the global economy to the detriment of the developing countries. In negotiations over trade, the countries of the industrial core had considerable advantages which worked in their favour. They had better trained government departments, with superior access to information, compared with their counterparts in the developing countries. In addition, the industrial countries could play the large number of small, weak developing countries off against each other and win greater trade concessions. They also dominated the global capital market and thereby secured preferential access to investment capital.

Further evidence of the disadvantageous trading position of the developing countries was seen in the apportionment of trade-related services such as shipping, insurance and banking among trading partners. Most developing countries were largely excluded from the services created by world trade. In contrast, such services were divided between the industrialized countries to the mutual benefit of both trading partners. The industrial countries also monopolized global research and development into technology: more than 99 per cent of technical research was undertaken in the industrial countries and reflected their requirements and interests.

Prebisch argued that, as a consequence of such forces, a given advance in productivity in the industrial countries tended to accrue to the owners of capital and to trade unions in those countries. It accrued to them in terms of higher profits and wages which might appear justified because of the production of a 'more advanced product'. By contrast, producers and workers in the developing countries lost out to the consumers of primary products (mainly those in the industrial countries), who captured the benefit of technical advance in terms of lower prices. For example, over the period 1930–65, the world sugar price rose four-fold, but the cost of the machinery imported from the industrial core for use in the sugar factories rose eight-fold.

The solution to such an unequal system of exchange was seen by the Prebisch school as being the rapid industrialization of the developing economies behind protective tariff barriers, i.e. the strategy of import substitution. Once domestic manufacturing capability had been built up, then markets would be opened to international trade. Meanwhile, to capture the scale economies of industrial production, developing countries needed to set up regional common markets with uniform external tariffs. The large scale-sensitive plants (like steel and automobiles), whose minimum optimum size outstripped demand in individual developing country markets, could be apportioned among the common market members (Gwynne 1985).

The dependency theory critique

The reforms to the international economy proposed by Prebisch and others were considered inadequate by the more radical dependency critique which emerged in Latin American in the 1960s. The dependency theorists dismissed the reforms proposed by the diffusionists. They argued that the development and expansion of the industrial core economy determines the performance of the dependent peripheral economy. Consequently, whereas the core can achieve self-sustaining economic growth, the dependent economies cannot.

Dependency theory: plantation society

Beckford (1972) applied dependency theory to the plantation economies, which he saw as comprising:

- the islands of the Caribbean and Indian Ocean;
- the Asian countries of Ceylon, Indonesia, the Philippines and Malaysia;
- the former African colonies of Portugal and Belgium, along with Cameroon and Liberia;
- the plantation 'sub-economies' of north-eastern Brazil and the US South.

Such economies are dependent on the core for markets, capital and technology. If the core finds a superior source of supply, then the initial developing country supplier may experience economic stagnation and lapse into a state of dependent underdevelopment.

Basically, the dependency theorists argue that the core distorts the allocation of capital, land and labour within the dependent economy, away from local peasant farmers and industrialists in favour of the often expatriate plantation owners. The resulting accumulation of capital may be ploughed back into the dependent economy, or recycled to the industrial core for investment there, or reinvested in new peripheral regions, whose expansion may prove detrimental to the original plantation economy. Beckford also stresses the crippling psychological legacy of dependency, in which local (coloured) people seek vainly to emulate the values of the metropolitan core within a racially stratified society which values lighter colouring.

While dependency theory usefully emphasizes the political economy of global relations, it has been weaker in providing hard empirical evidence to support its conclusions (Seers 1981). For example, Beckford's critique of plantation society in the Caribbean had lost much of its force by the time he published his book in 1972. This is because postwar changes in political power had seen the emergence of trade unions and universal suffrage in the Caribbean, which shifted the balance of political forces strongly in favour of local interests. There is evidence that, whereas the dependency critique certainly held for slave society (Higman 1976) and the nineteenth-century period of indentured Asian labour (Adamson 1972), the seeds of a new plantation system which was much more sensitive to local economic interests than the old paternalistic system were already emerging by the 1930s (Auty 1975).

There is evidence that in the Caribbean, during the postwar years the newly strengthened local institutions such as the trade unions became too successful in redistributing the returns from the sugar corporations towards workers and peasant farmers. The large MNC-owned plantations could not make sufficient profit on sugar to yield an adequate return on their investment, so that under-investment occurred in both the factories and field mechanization (Auty 1976). With insufficient improvements in productivity the Caribbean region ceased to be a competitive sugar producer and struggled to maintain production, even with subsidized prices in the EC market.

For example, Jamaican sugar production costs were around $430 per tonne in the late-1980s, compared with world prices of $230 per tonne and a target price in the protected EC market of $290 per tonne. By the 1980s, most Caribbean sugar cane plantations had been nationalized, a change which merely accelerated the decline in their efficiency, for reasons discussed later in this chapter. Labour problems mounted and by the 1980s Jamaican output was less than half of its 1965 peak level.

Divergence within dependency theory

As dependency theory evolved, it exhibited a widening range of interpretations. For example, Walton (1975) emphasized internal colonialism, which down-plays the role of external agents and stresses the importance of local elites in the exploitation of the peripheral economy. This is in direct contrast to Dos Santos (1969), who identified MNC dominance as a new form of dependency in the postwar period, the latest in a historical sequence. In the context of Latin America, that sequence ran from colonial dependency under Spain and Portugal, through the financial and industrial dependency of the late-nineteenth-century British bankers and into the technological-industrial dependency of the mainly US MNCs in postwar years.

Under colonial dependency, the imperial countries of the core monopolized trade, land, labour and manpower in the colonies. This system was replaced by the financial-industrial dependency of the late nineteenth century, in which the 'hegemonic' centres of north-west Europe and the north-eastern United States dominated the accumulation of capital. They invested in the periphery for raw materials and products to be consumed in the industrial core countries. They built railways which opened up the peripheral economies to industrial competition as well as to export opportunities.

Dos Santos argued that the postwar technological–industrial dependency was spearheaded by the MNCs. Contrary to the depiction of 1950–73 as a golden age, Dos Santos considered that the diffusion of economic growth failed after the Korean War. This was because 'monopoly capitalism' (reflecting imperfect markets or oligopolistic competition, which the MNCs exploited to their advantage) strengthened its hold on the periphery. The Latin American countries' unfavourable terms of trade had constrained their ability to import the capital goods which they needed to build up their industrial sectors. They had imported manufactured goods at inflated prices which included capital remittances, royalty payments on technical assistance and high freight charges which were to the benefit of the industrial countries.

In addition, the debt-service on such foreign capital had absorbed scarce foreign exchange. The situation became chronic with the onset of the debt crisis in 1982, when as much as 6 per cent of GDP was being transferred overseas annually in debt service. Previously, the developing countries had been net capital importers at a rate equivalent to 1–5 per cent of GDP (Syrquin and Chenery 1989).

Dependency theorists concluded that the LIEO rendered the developing countries powerless to achieve the political and economic reforms hoped for in the 1950s. For them, the failure of the reformist solutions advanced by Prebisch justified more radical measures which challenged the entire postwar economic system that was founded on the Bretton Woods convention. The brunt of their attack was aimed at the MNCs, especially those like plantations and mining firms, whose control of sizeable tracts of land and the resources they contained led to their depiction as a 'state within the state' (Porteous 1973). This hostility found expression in the wave of postwar nationalization which peaked in the early-1970s (Fig. 8.1).

The multinational corporation as scapegoat

The rise of the MNC

Estimates made in the early-1970s suggested that the MNCs accounted for 15 per cent of GWP – some $450 billion – and had been growing at 10 per cent annually since the early-1950s, twice as fast as the global economy. Projecting the 1950–70 trend to the year 2000 revealed that the MNCs might easily account for 50 per cent of GWP by then (Bergman 1971). Several MNCs already dwarfed many nation states in terms of economic size: in 1971 the top ten MNCs each generated more added value than the GDPs of eighty nations. There was fear that such unequal economic power would lead to a loss of sovereignty by the developing countries to the giant corporations.

The ownership of MNCs was dominated at that time by the United States: two-fifths of MNC investment was made by US corporations abroad, while a further one-fifth of MNC investment comprised foreign investment within the USA. US firms dominated the foreign investment of some developing countries, including Mexico, Venezuela, Argentina, Chile and Saudi Arabia. The UK was the next largest source of MNC investment, with 14 per cent of the world total, and more than half the foreign investment in its former colonies of India and Nigeria. Anti-MNC rhetoric was used by some developing country politicians to make MNCs the scapegoats for the disappointing postwar economic performance of their countries. For example, many blamed the seven oil majors (Exxon, Shell, BP, Texaco, Mobil, Chevron and Gulf) for orchestrating the large price increases which occurred during 1973–4, although in reality the oil majors were considerably weakened by the oil shocks.

Capital deepening versus capital widening

Hymer (1976) developed an influential theoretical critique of the MNCs. He argued that private corporations had largely solved the problems of coordinating large business enterprises by the 1920s, so that scale no longer acted as a constraint on their expansion. They had achieved this by splitting the organization of the company into three units: the head office, the division and the field unit. The function of the head office was to monitor long-term strategy decisions and to allocate capital to those divisions which could achieve the highest return. The divisions handled actual operations and were organized on the basis of either product lines or geographical areas (Rumelt 1986). Finally, the field unit (i.e. the mine or factory producing the product) was to function as a cost centre.

Such an organizational system overcame the potential diseconomies of scale of large bureaucratic organizations and, according to Hymer, gave the MNCs two alternative strategies. They could opt either for capital widening, or for capital deepening. Capital widening involved saturating the existing industrial country markets with known technology and products and then shifting investment to sell those same products in new developing country markets as the returns in the core declined. Such a process of capital widening would eventually lead to a globally even pattern of income distribution.

In contrast to capital widening, capital deepening would amplify the global income gap. This is because capital deepening constantly introduces new technology and/or products into established (industrial country) markets. This boosts the productivity and incomes of the workforce in such markets and postpones the onset of diminishing returns, so that the incentive for MNCs to seek new markets elsewhere is reduced.

Hymer argues that in opting for capital deepening, the MNCs sharply widened global income differentials from the 1930s. His starting premise finds some support from Krugman (1979) who, however, draws quite a different conclusion. Krugman concludes that the industrial countries do need to introduce a constant stream of new products, but not simply to advance their per capita incomes. Rather, they need to develop new products in order to stop their incomes falling back as the newly industrializing countries (NICs) capture an increasing share of the more labour-intensive and even skill-intensive manufactured goods. Certainly, as Chapter 11 shows, the product cycle model indicates that competitive advantage in many scale-sensitive manufactured goods such as steel and mass-market cars (and not simply in labour-intensive goods like textiles) began passing to the larger NICs in the 1980s.

Beenstock (1983) goes so far as to blame the post-1973 slowdown in the world economy on an overly slow response of the industrial countries to the emerging NIC challenge. The failure of the industrial countries to restructure their economies to produce more advanced products was reflected in a decline in the return on capital in the industrial countries, and the 1970s' surge of capital to the rapidly developing NICs. Far from the MNCs widening global income differentials, as Hymer argues, they are seen in this view as a potent force for narrowing them. This is because of the ability of MNCs to transfer capital and technology to places where it makes the highest return.

But even as the LIEO was bringing such trends about, the anti-MNC rhetoric of the dependency theorists and others caused many developing country governments in the 1970s to accelerate their progress along a painful interventionist learning curve. The problems which they encountered can be illustrated with reference to the application of dependency theory to the mineral economies. In the 1960s, the ownership of mines, oilfields and plantations in the developing countries was still dominated by MNCs.

Girvan's mineral economy thesis

Girvan (1971) applied dependency theory to bauxite mining in the Caribbean, with particular reference to Jamaica. He calculated that between 1955 and 1965 Jamaica retained less than 3.5 per cent of the value added along the aluminium production chain (Table 8.2). This was because the three processing stages (refining the bauxite into alumina, the smelting of alumina and the fabrication of aluminium ingot) mainly took place in the industrial countries. Girvan attributed this to the dominance of the global aluminium market by an oligopoly of six MNCs (Alcan, Alcoa, Reynolds, Kaiser, Pechiney and Alusuisse) which were responsible for two-thirds of global aluminium production. He argued that oligopoly facilitated price collusion among producers whose products were virtually identical in quality.

Control of the market freed the MNCs from locational constraints, according to Girvan. The MNCs were able to locate smelting and fabrication (which comprise three-quarters of the added value in the bauxite production chain) in 'safe' but economically sub-optimal locations in the industrial countries. The absence of refineries, smelters and fabricating plants in the country where the bauxite was mined explains why Jamaica received so little benefit from its natural resource. Worse, the mines were supplied with inputs such as chemicals from industrial country sources, rather than from Jamaican factories, further muting the local multiplier from mining.

Yet if a developing country government wished to capture more of the potential benefits from mining through higher taxation, the MNC could use transfer pricing to escape the tax. The MNCs were free to set prices at whatever levels minimized their tax liabilities, because they sold most of their production among their own subsidiaries within their vertically integrated

Table 8.2 *Inputs used and value added along the Jamaican bauxite chain (£ million)*

Process stage	In Jamaica		Abroad		Total	
	Input	Value added	Input	Value added	Input	Value added
Bauxite mining	61	167	–	–	61	167
Alumina refining	52	96	196	365	248	461
Aluminium smelting	–	–	984	1481	984	1481
Aluminium fabrication	–	–	1839	2908	1839	2908
Total	113	263	3019	4754	3132	4754
Percent total value	1.3	3.2	37.0	58.3	38.3	61.5

Source: Girvan (1971)

production chains. The MNCs could evade all tax liability by simply equating the price and cost of bauxite, for example, and declaring no profit in the mining country.

For Girvan, the obvious answer to such abuses was nationalization of the mines, and this was what the Guyanese government did in 1971. The Jamaican government was more cautious, however, and opted to increase its equity in the MNC subsidiaries, while also introducing an OPEC-style tax on bauxite in 1974. The tax boosted bauxite revenues to 6 per cent of Jamaican GDP (Table 8.3). The move was made as part of the International Bauxite Association cartel, which sought to maintain a uniform delivered price at the principal markets. This meant that those producer countries which enjoyed rents, either as a result of mining relatively cheap bauxite (like Australia) or through low freight rates (like Jamaica to the United States), could capture the benefits, rather than have them accrue to the aluminium MNCs and their customers – provided the cartel held together.

But Girvan neglected the fact that if the MNCs did not compete on price, they competed on cost. This meant that the MNCs were sensitive to the changing competitiveness of individual countries. The MNCs calculated that the Jamaican bauxite tax over-estimated the rent on Jamaican bauxite by at least one-third. During the global recession of the mid-1970s, the MNCs concentrated their cutbacks on high-cost operations like Jamaica; the imposition of the bauxite tax was associated with a halving in Jamaican production (Table 8.3).

The bauxite tax accelerated the long-term loss of Jamaica's global bauxite market share, which fell from 20 per cent in the early 1970s, to 7 per cent by the late-1980s. The long-term decline reflected two sets of factors: first, intensified competition from newer bauxite producers like Guinea and Australia, and second, the maturation (i.e. deceleration) of aluminium demand in Jamaica's principal markets (the southern United States and Europe). In

Table 8.3 *Trends in Jamaican bauxite production and revenue, 1974–85*

	1974	1979	1984	1985
Bauxite production (million tonnes)	15.3	11.6	6.5	9.9
Alumina production (million tonnes)	2.8	2.1	1.8	2.2
Mining as % total exports	55.0	71.4	63.1	58.7
Levy as % GDP	6.1	7.2	5.1	2.1
Levy as % total tax revenue	n.a.	25.8	17.9	4.6

Source: Auty (1993a)

effect, the bauxite tax made Jamaica a high-cost source, and it became a swing producer of bauxite and alumina. In other words, a substantial fraction of Jamaican productive capacity was used to meet cyclical peak demand, and was shut down when that demand declined.

The initial results, however, were highly favourable for Jamaica, as was also the case with government intervention to promote populist booms in Latin America (Chapter 6). The bauxite tax boosted the sector's contribution to 22 per cent of total government revenues during 1974–6. The Jamaican government wisely set up a capital fund to purchase shares in the aluminium sector, but most of the resources leaked swiftly into current (day-to-day) expenditure. Then, as bauxite production fell, the contribution of the tax to GDP dropped to 2 per cent by 1985 (Table 8.3) – barely one-third of what it had been expected to yield when it had been imposed. The tax was reformed in 1988: it was halved and charged to production costs. The MNCs then paid 33 per cent corporate tax on their profits, so that the principal weakness of the tax – its inadequate link to corporate profitability – was removed.

The importance of mining to the economy meant that Jamaica took 2 decades to recover from the mismanagement of its bauxite rents. Its real GDP declined by 1 per cent annually after the imposition of the tax, and per capita incomes fell by even more. Sharpley (1983) estimates that if pre-tax economic trends had continued, Jamaicans would have been twice as wealthy as they actually were by the early 1980s. Robinson and Schmitz (1989) suggest that sustained economic recovery from the bauxite-triggered recession of the mid-1970s finally began in 1985. But Jamaican GDP growth averaged only 1.0 per cent during 1982–9, not enough to prevent further per capita income decline.

The weak performance of the Jamaican economy in the late 1980s is the more disappointing because it occurred despite a rebound of the bauxite sector and a sustained high rate of investment in the national economy. The feeble economic performance reflected the neglect of the Jamaican non-mining tradables such as sugar, bananas and manufactured goods. This neglect was in turn linked to over-optimistic expectations of the contribution from bauxite mining, a common problem of mineral economies which is discussed in Chapter 9. Meanwhile, many mineral economy governments which nationalized their mines fared little better.

State-owned enterprise performance

Optimistic observers like Radetzki (1985) argue that nationalization initially weakens enterprise performance because the new state firm lacks the experience of the MNCs. But a 10–20-year learning curve improves performance and pushes it back towards MNC levels of efficiency (Fig. 8.2). Radetzki considered such an outcome quite satisfactory, because the social benefits of state ownership more than offset any modest efficiency losses. In fact, Radetzki's view was based on only three case studies, one of which conformed to his model.

The more usual experience has been for a relatively smooth state takeover with several years of apparently satisfactory operation (Auty 1986). But productive efficiency is steadily eroded because equipment maintenance is skimped in order to show a satisfactory profit and to attract the large capital investments which mining requires. Meanwhile, inadequate managerial autonomy caused by political interference in decision-making results in an accelerating turn-over of managers, who are all too often replaced with political nominees rather than technocrats. Labour force cooperation also declines, as workers strike to test their ability to tap the perceived unlimited financial resources of the state. The net effect is a fundamental weakening of state enterprise performance, which is abruptly revealed by some price or cost shock which causes the industry to experience an accelerating decline in output and profitability (Fig. 8.2).

Shafer (1983) has argued that the weakness of many state enterprises in developing countries reflects the fact that too many governments sacrifice the commercial efficiency of state firms in order to buy political support from strong pressure groups. In addition, government ownership of firms creates rents (returns above a normal profit) because of the perceived low risk of bankruptcy. The rents are shared between workers and politicians at the expense of lost export markets and, ultimately, of the taxpayer, who must subsidize the losses of the state firm. Shafer concludes that the dependency theorists underestimate the dual role of the MNC as an insulator. First, MNC ownership protects firms from the debilitating predatory interference of politicians anxious to use their control of the state enterprise to buy political support. Second, it promotes global competitiveness, by assuring the firm has adequate access to global markets, technology and capital.

With hindsight, the dependency theorists overstated the power of the MNCs. In fact, MNCs provide only one-eighth of the foreign capital flowing into the developing countries (Lewis 1978). Such foreign capital is, in turn, itself only one-fifth of the total capital invested by the developing countries.

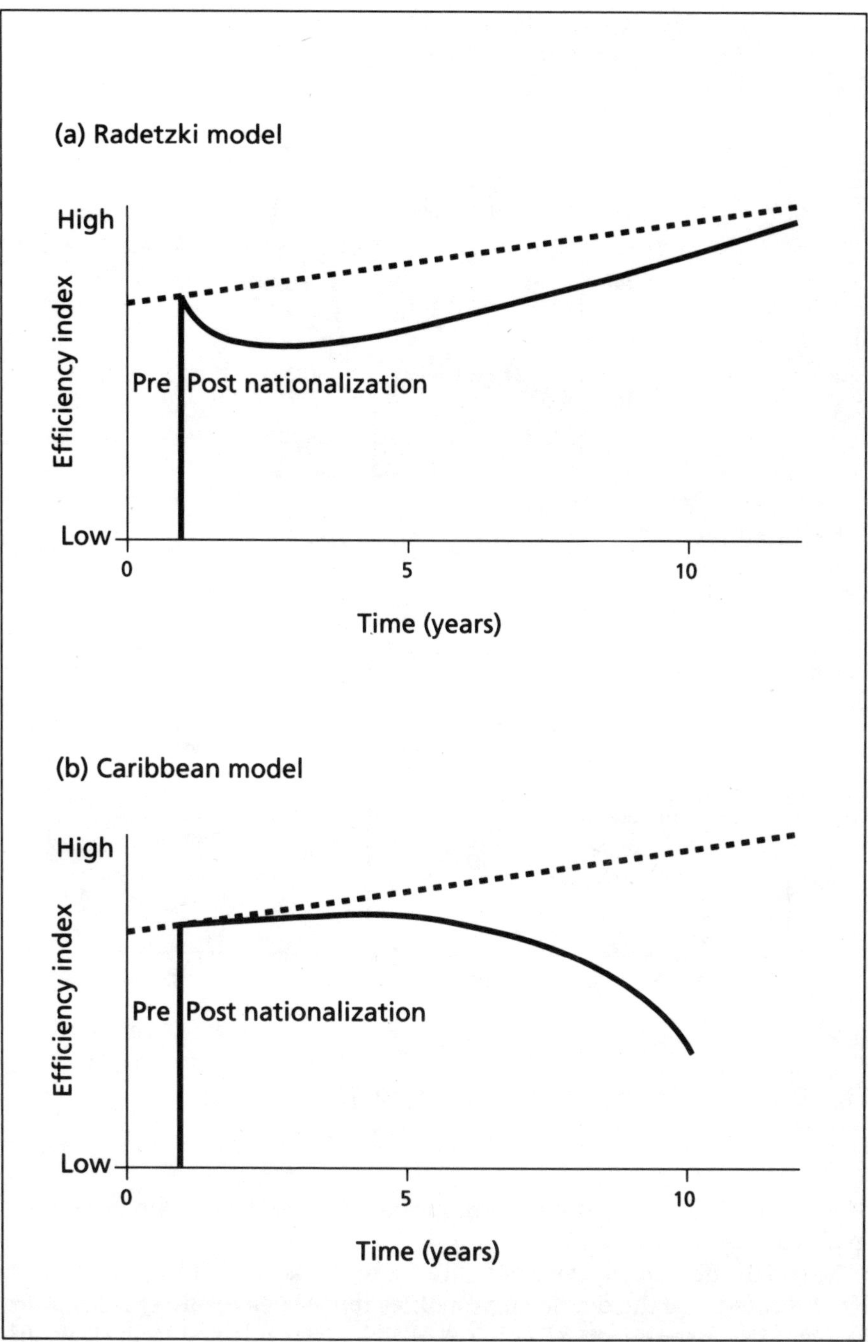

Fig. 8.2 *Impact of nationalization on enterprise efficiency*

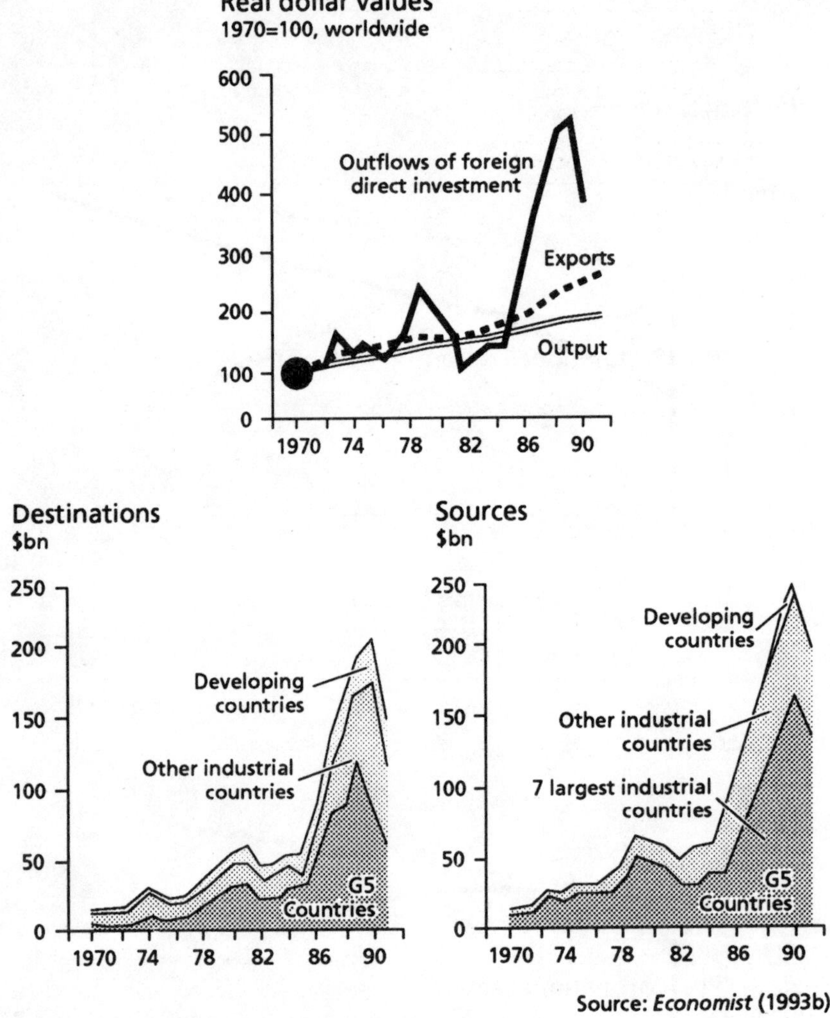

Fig. 8.3 *MNC investment flows, 1970–90*

This means that the MNCs account for only 2.5 per cent of the investment in developing countries.

Nor do the developing countries attract a large fraction of MNC investment. Fig. 8.3 shows that the developing countries represent a relatively small share of total MNC investment, 83 per cent of which was oriented to the industrialized countries by the late-1980s—up from two-thirds in the early-1970s (Dunning 1993, UN 1992). MNC investment in developing countries shows quite distinct regional preferences among the three leading sources (Fig. 8.4).

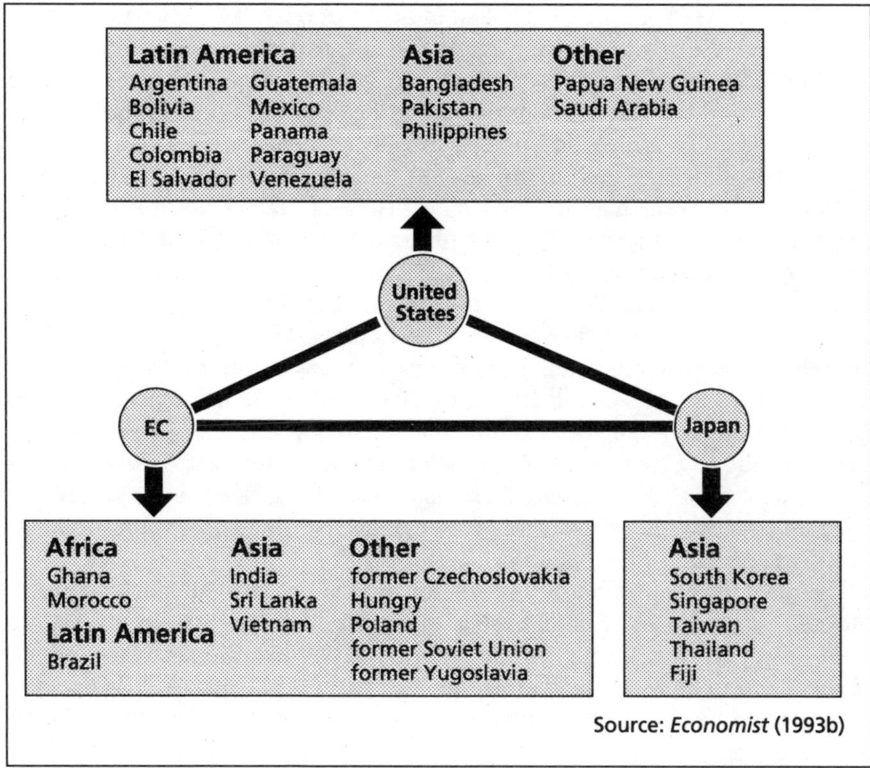

Fig. 8.4 *MNC investment clusters*

Finally, most MNCs are relatively small firms: for example, 78 per cent of UK MNCs employ fewer than 500 workers and 50 per cent of global firms are classified as small or medium. The so-called 'monopoly' capitalists such as the seven oil majors and six aluminium sisters have steadily lost market share throughout the postwar period. Moreover, due to tight government supervision, they do not appear to have earned profits in excess of the manufacturing average.

State intervention as the cause of poor economic performance

Balassa (1988) and other orthodox economists have argued that the poor economic performance of the developing countries during the 1970s was due to excessive levels of government intervention. This manifested itself in the adoption of inward-oriented trade and industry policies, an expansion in state ownership of directly productive assets through nationalization, and efforts to manipulate commodity prices through producer country cartels. Studies by Balassa and others for the World Bank built up a wealth of empirical data through the 1970s which both Prebisch and the dependency groups lacked. The studies underline the weakness of inward-oriented policies (World Bank 1993a), the inherent bias towards inefficiency of state enterprises (Shirley 1983) and the superiority of export diversification over the manipulation of commodity prices by cartels (Duncan 1993).

Lal (1983) has made an even more pointed attack on the role of state intervention as a barrier to rapid economic development. He argues that, contrary to Prebisch, developing country governments must let the market play more of a role, not less. Meanwhile, he notes that the share of manufactures in the developing country non-fuel exports (i.e. excluding oil) was 44 per cent by 1978, and above 75 per cent in the case of the successful East Asian countries like Taiwan, South Korea, Singapore and Hong Kong. Even in large inward-oriented countries like Brazil and Mexico, manufactured exports rose to more than 50 per cent of total exports.

The main developing country recipients of MNC investment, such as Brazil, Mexico, Singapore and Korea, have always controlled it tightly. Ironically, the main error with regard to MNCs is to do just what Prebisch and others advised, and to limit competition within the markets in which MNCs operate. By providing the MNCs with a closed market through the imposition of high tariff barriers, developing governments sanction the MNCs to cream off the resulting rents (returns in excess of a normal profit). The MNCs are unconstrained by either global or domestic competition in such cases.

In contrast, where MNCs are forced to function in competitive markets, they generate foreign exchange, create employment and diffuse skills. By the 1990s the developing countries sought to attract MNC investment. The MNCs provide 65 million direct jobs world-wide, some 3 per cent of the world labour force. In the case of Singapore, one of the most successful NICs, the MNCs account for 55 per cent of that very prosperous country's manufacturing jobs, 63 per cent of its output and 90 per cent of its total exports (*Economist* 1988).

Summarizing, even as influential commentators despaired of the developing countries achieving rapid industrialization, such a process was under way. And it was occurring especially rapidly in the resource-deficient countries which were relatively open to international trade. Consequently, the attempts to construct a new international economic order were an unnecessary and costly diversion. Few countries illustrate more clearly the hopes, opportunities and disappointments of such attempts than the mineral economies, as Chapter 9 shows.

9

The mineral economies: from trade confrontation to sustainable development

This chapter first outlines the main characteristics of the mineral economies, recognizing two main groups: the oil-exporters and the ore-exporters. The second section of the chapter then contrasts the use of the oil-windfalls (rents from higher oil prices) associated with the 1973 and 1979 oil shocks by two oil-exporters, Saudi Arabia and Nigeria. The third section examines the impact of the prolonged post-1973 downswing in mineral ore prices (copper and bauxite, for example) on the ore-exporting countries.

The final section of the chapter sets out the policies required for the sustainable development of mineral-driven economies. Briefly, sustainable mineral-driven development requires two conditions. First, provision must be made to replace the depleting mineral resource with some other income-generating asset. Second, the environmental degradation associated with mineral extraction and use must be minimized.

The mineral economies

The mineral economies emerged as an important group of developing countries in the 1970s, when fundamental changes occurred in world mineral consumption. These changes resulted from an unexpected slowdown in global mineral demand, which resulted in greater volatility in mineral prices (Fig. 9.1). The mineral economies are defined by Nankani (1979) as those developing countries which generate at least 40 per cent of their exports and 10 per cent of their GDP from minerals (Table 9.1). They comprise about one-quarter of all developing countries and have three advantages for economic development over countries which lack the mineral bonus. First, mineral exports earn extra foreign exchange. Second, they are an additional source of government revenues in addition to those furnished by agriculture. Third, the processing of minerals into finished products provides an added industrialization option along with import substitution and competitive exports.

Table 9.1 *Economic and social characteristics of prominent mineral economies*

Country	Area (million sq. km)	Population (million)	Cropland/hd (ha)	Population growth (%/yr)	Population in cities 1991 (%)	GDP growth 1980–91 (%/yr)	Per capita income ($1991)	Literacy (%)	Life expectancy (yrs)
Oil exporters									
Cameroon	0.475	11.9	0.59	2.8	42	1.4	850	56	55
Ecuador	0.284	10.8	0.25	2.6	57	2.1	1000	86	66
Indonesia	1.905	181.3	0.12	1.8	31	5.6	340	51	52
Nigeria	0.924	99.0	0.29	3.0	36	1.9	340	51	52
Saudi Arabia	2.150	15.4	0.08	4.6	78	(0.2)	7820	62	69
Venezuela	0.912	19.8	0.20	2.6	85	1.5	2730	88	70
Ore exporters									
Bolivia	1.099	7.3	0.47	2.5	52	0.3	650	77	59
Chile	0.757	13.4	0.34	1.7	86	3.6	2160	93	72
Jamaica	0.011	2.4	0.11	1.0	53	1.6	1380	98	73
Morocco	0.447	25.7	0.37	2.6	49	4.2	1030	49	63
PNG	0.463	4.0	0.10	2.3	16	2.0	830	52	56
Peru	1.285	21.9	0.17	2.2	71	(0.4)	1070	85	64
Zambia	0.753	8.3	0.62	3.6	51	0.8	460	73	49

Source: World Bank (1993b), WRI (1992)
Note: Other mineral economies include Algeria, Botswana, Brunei, Guinea, Guyana, Iran, Iraq, Kuwait, Libya, Namibia, Surinam, Trinidad and Tobago, Zaire, Gabon, Congo, Jordan, Mauritania, Niger, Togo

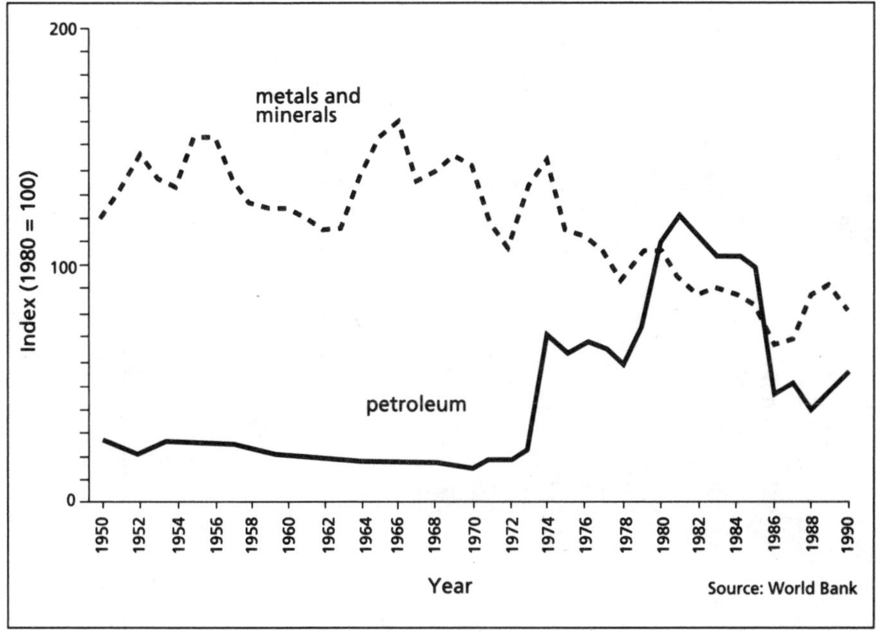

Fig. 9.1 *Oil and metals prices, 1950–90*

Despite these potential advantages, the mineral economies have seldom outperformed other groups of developing countries (Table 9.2), and the achievements of individual mineral economies have been erratic (Roemer 1985, Wheeler 1984). During colonial times, many MNC mining subsidiaries functioned as economic enclaves, with a minimal local economic impact and maximum growth stimulus to distant metropolitan regions (Kessel 1977, Thoburn 1977), as dependency theorists like Girvan (1971) argued. Chapter 8 noted that the capture of a larger share of the rents (that is, returns in excess of those needed by an efficient producer to remain competitive) from the mining subsidiaries of MNCs was an early post-independence goal of many mineral economies. Success in that area, however, brought unforeseen problems of making effective use of the additional resources for development.

The management of commodity booms has proved especially difficult for mineral economies. Far from strengthening the economy, the domestic spending of mineral windfalls risks overburdening the public sector and weakening the competitiveness of the non-mining traded sectors. The 'tradable' sectors (mainly agriculture and manufacturing) are those exposed by trade to international competition, and exclude most services. Prudent windfall deployment requires that during the booms, governments accumulate financial reserves for two reasons. First, to avoid outstripping the absorptive capacity (that is, the available infrastructure and skilled labour) of the

Table 9.2 *Investment and growth rates by developing country group*

Measure		Hard-mineral exporters 1960–71	1971–83	Oil exporters 1960–71	1971–83	Other middle-income countries 1960–71	1971–83	Other low-income countries 1960–71	1971–83
Investment/GDP:	Mean	0.21	0.23	0.21	0.28	0.20	0.24	0.14	0.17
	SD	0.06	0.05	0.10	0.10	0.05	0.05	0.04	0.06
Gross IOCR[a]:	Mean	0.28	0.07	0.34	0.12	0.32	0.17	0.26	0.17
	SD	0.05	0.02	0.06	0.05	0.02	0.01	0.03	0.04
Growth of GDP	Mean	2.5	–1.0	2.9	1.9	3.7	2.0	1.3	0.7
per capita (per cent)	SD	1.1	1.2	1.7	3.7	1.8	2.3	1.4	2.2
Number of countries		10	10	10	10	29	29	20	20
Terms of trade indices	Metals and hard minerals		Petroleum		Agriculture				
(relative to unit value of manufactures imported by developing countries)									
1960–2	100		100		100				
1970–2	104		92		91				
1980–2	78		636		84				

[a]Incremental output/capital ratio
Source: Gelb (1988)

economy, and second, to smooth the adjustment of the economy to the post-boom downswing.

If the windfall absorption proceeds too quickly, then the price of labour and land will rise. Such inflation pushes up the real exchange rate, so that domestically produced agricultural and manufactured goods become less competitive, and unless they are given subsidies, their share of GDP shrinks. This phenomenon is known as 'Dutch disease'. The agricultural sector of mineral economies is typically one-half or less of the norm for countries of a similar size and level of development. Manufacturing would also be smaller than the norm if it was not protected by high tariffs. But such weak and/or shrunken manufacturing and farm sectors find it difficult to expand when mineral revenues decline. Economic growth is therefore erratic, with faster growth during booms rarely compensating for the post-boom slowdown.

Corden (1982) traced 'Dutch disease' to the operation of two effects: the short-term windfall spending effect and the more long-term resource movement effect. During a mineral boom the domestic spending effect of the mineral windfall causes the price of protected non-traded goods (mainly services) to rise relative to the prices of traded goods, whether imported or domestically produced. This is because the tradables are subject to price restraint by way of global competition. In this way, the exchange rate is overvalued, and agriculture and manufacturing have difficulty competing with imports. The second effect, long-term resource movement, is triggered by the prospect of higher returns in service and mining sector businesses. It involves a shift of labour and capital out of the lagging traded sector (agriculture and manufacturing), causing its growth to slow further or even to contract. The resulting weakening of the traded sector leaves the economy vulnerable when the mineral boom ends and alternative sources of foreign exchange and taxes are needed.

As noted earlier, one important reason for the government to accumulate savings during a commodity boom is to mute the 'Dutch disease' effect, but a second is to smooth the subsequent adjustment to post-boom conditions. Lewis (1982) examined the tendency of developing country governments to misallocate boom-time spending and investment. He found that most governments do not accumulate the savings abroad during the boom which are required in order to reduce the risk of Dutch disease and also to smooth the adjustment to the post-boom downswing. Most governments find it difficult to resist political pressure to use the windfalls for immediate consumption and pork-barrel investment programmes (Gelb 1988).

Instead of saving an appropriate fraction of the windfall, the governments of mineral economies invariably increase consumption by, for example, subsidizing fuel prices or mortgages. This creates patterns of expenditure that are unsustainable during the downswing, but which are nevertheless difficult to rein in. Meanwhile, public sector management is overstretched during boom conditions, so that harassed civil servants favour investment in a few large, capital-intensive projects, many of them ill-conceived. During post-boom downswings such projects may not generate sufficient revenues to repay the loans with which they were constructed, let alone to compensate for the reduction in mineral revenues. They therefore exacerbate the adjustment to the loss of mineral revenues, as the boom wanes (Auty 1990a).

Such cyclical adjustments create problems for all primary product exporters, but they tend to be more acute in mineral economies for several reasons. First, the capital-intensity of mining initially concentrates the windfall on the mine owners in the form of higher profits (which the government must tax away without depressing the incentive to invest) and upon a small highly paid labour aristocracy. Second, for reasons noted above, the government may fail to spend the mineral windfall in a way which advances national welfare. Third, high mining sector wages have a demonstration effect which leads to subsidies and wage increases in the non-mining sector that are unrelated to productivity and which are also unsustainable through the long term. Finally, this 'spread effect' from the mine workforce does not equalize social opportunities: the mineral economies tend to have a more skewed income distribution and also lower literacy rates than their non-mineral counterparts (Nankani 1979), making for heightened political tension.

But some mineral economies cope more effectively with mining booms than others. One important determinant of success is the nature and size of the mineral resource. For example, the mirror image of the oil exporters' booms has been the downswing problems of the hard mineral exporters of copper and bauxite (Table 9.1). But the main difference in performance derives from the policy adopted. A comparison of mineral-exporting countries shows which policies minimize the risk of the damaging boom-time effects and thereby enhance economic performance. It also shows that the large transfers of capital from rich to poor countries which higher commodity prices bring (and which Sakharov and Snow advocate via aid) are likely to be wasted because of inadequate domestic absorptive capacity.

The oil-exporting countries

OPEC and oil price volatility

In 1980 the Organization of Petroleum Exporting Countries (OPEC), whose members accounted for 350 million people and 75 per cent of global oil reserves, was at the height of its power. If the non-OPEC oil exporters are added, then the oil-exporting countries as a whole embraced 510 million people, with an average per capita income of $1400. The oil-exporting countries can be usefully subdivided into the 'low absorbers' and 'high absorbers', and therein lies a critical source of friction in setting OPEC's pricing strategy.

The low absorbing oil-exporters are sometimes described as the capital-surplus oil exporters. These are countries whose oil reserves are large in relation to their population, so that their oil revenues greatly exceed the present capacity of their domestic economies to invest and consume them effectively. Iraq, Libya, Kuwait and Saudi Arabia fall into this category. Their combined population in 1980 was 27 million and their per capita income averaged $5000. The low absorbers' long-term interest within the OPEC cartel lay in preventing sharp increases in oil prices, because that would accelerate the industrial countries' substitution of oil by alternative energy sources such as nuclear or solar power (Mesarovic and Pestel 1974).

In contrast, the high absorbing oil-exporters have large populations in relation to their oil reserves, so that the scope for immediate domestic consumption and investment of the oil revenues may outstrip the revenues available. The combined population of this group was 480 million in 1980 and its per capita income averaged only one-fifth that of the low absorbing countries. Important countries within this group include Indonesia, Nigeria, and Venezuela (Table 9.1). Their conventional reserves of oil (as opposed to the non-conventional sources like the Orinoco heavy oils) are believed to be nearer to exhaustion than the low absorbers' reserves, so that the high absorbers have less concern over the loss of oil markets in the future. Their main objective has been to maximize oil revenue now and to transform it into productive assets as quickly as possible for the time when their oil reserves are exhausted.

The oil supply disruptions associated first with the 1973 Arab–Israeli war, and second with the overthrow of the Shah of Iran in 1978, both created an opportunity for OPEC to raise oil prices. The cost of a barrel of oil rose in real terms from $2.30 to $5.50 between 1974 and 1978, and then to $11.50 during 1978–81 (Fig. 9.1). Each of these shocks was positive for the oil-exporters and transferred the equivalent of 2 per cent of GWP to them from the oil importers. However, the unexpectedly powerful impetus to conservation which they engendered among the oil importers depressed world oil demand. OPEC then failed to agree adequate production cutbacks and this caused oil prices to decline below $3 per barrel (in 1970 prices) by the early-1990s, with an especially sharp fall in 1986.

But before that, the positive shocks greatly increased the financial resources available to the oil-exporters. Measured in terms of their non-mining GDP, the annual windfall from higher oil prices among the high absorbers ranged from around 10 per cent of GDP in Venezuela to 20 per cent for Nigeria and almost 40 per cent in Trinidad and Tobago (Gelb 1988). In contrast, low absorbing Saudi Arabia's oil windfall approached 200 per cent of its GDP (Auty 1990a). Not surprisingly, the range in windfall size yielded very different options for the two subsets of oil-exporters.

The low absorbers could choose between becoming *rentier* economies (by accumulating sufficient overseas assets to be able to live comfortably on the interest and dividends), or investing in the domestic economy to diversify it

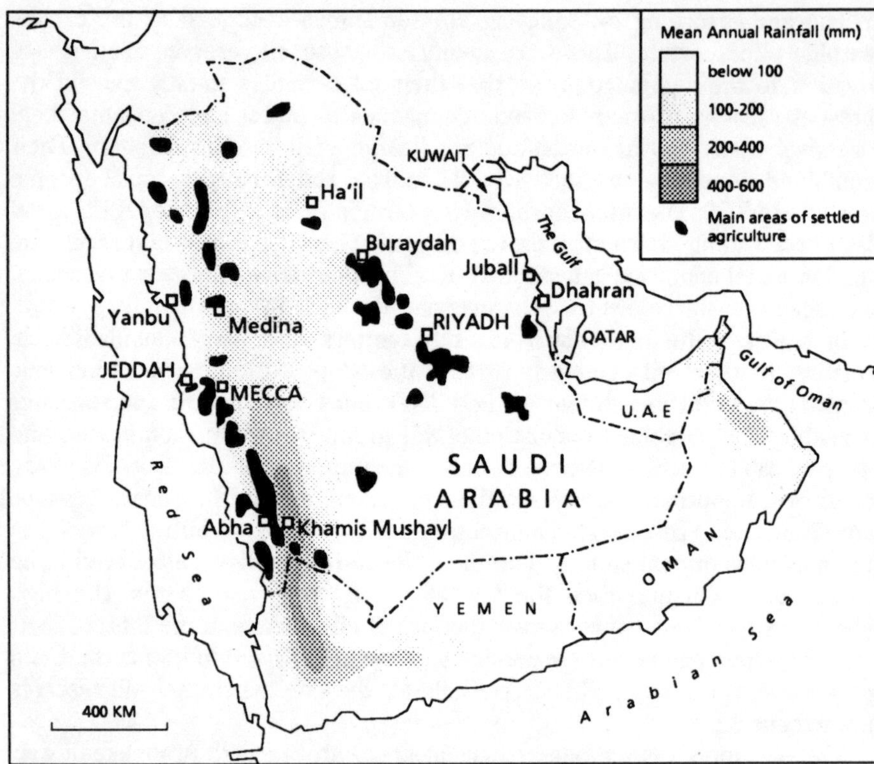

Fig. 9.2 *Saudi Arabian land use and industrial growth poles*

away from oil in anticipation of a time when oil might cease to be important. Kuwait favoured the former option. Stauffer (1985) estimates that by the mid-1980s, Kuwait relied on earnings from overseas investments for around half its annual needs, with domestic injections of oil revenues accounting for the rest. (Two-thirds of the country's overseas assets are believed to have been liquidated in the subsequent conflict with Iraq.) In contrast, Saudi Arabia opted to diversify its economy on the basis of resource-based industrialization (RBI), mainly the downstream processing of oil and gas into petrochemicals.

Saudi oil windfall deployment

The Saudi oil windfall was the equivalent of an extra 200 per cent of GDP annually during the first oil boom of 1974–8, and around 180 per cent of GDP during the second boom of 1979–81 (Auty 1990a). The Saudis saved around one-third of their windfall abroad in real estate, shares and bonds in the industrial countries, and by 1981 the country's total overseas assets topped $140 billion. Meanwhile, the government increased the country's capacity to absorb increased expenditure by opening up the economy to foreign labourers, overseas construction firms and imports of manufactured goods. This kept domestic inflation in check.

The Saudi government initially invested heavily in a once-for-all push to establish the infrastructure of a modern country. Expenditure on the construction of highways, ports, airports, hospitals and universities ran at $4 billion/month at its peak. In the early-1980s the emphasis shifted towards resource-based industry (RBI), with the investment of more than $50 billion in the construction of two heavy industry growth poles at Al Jubail and Yanbu (Fig. 9.2). The RBI comprised mostly petrochemicals and some steel, using cheap gas feedstocks to offset the country's higher costs of plant construction, labour and freight. The RBI plants were built efficiently, mostly as joint ventures between leading MNCs and Sabic, the state-owned holding company. The Saudi government then withdrew from the active promotion of industry and encouraged ongoing industrial diversification through private investment in petrochemical processing and consumer goods manufacture.

The results of the Saudi windfall deployment were impressive. GDP expanded almost three-fold between 1970 and 1985, averaging 8 per cent growth annually (Table 9.3). Non-oil output expanded even faster, averaging 12 per cent annually, and the share of oil in GDP dropped below half by 1983. Annual oil revenues peaked at $120 billion in 1981 before they levelled off at $25 billion in the late 1980s. But when oil prices fell sharply in the mid-1980s, the economy began to contract because the scale of diversification away from oil which was required was too great to be accomplished within a single decade.

The Saudi economy remained oil-driven despite a decade of diversification efforts (Table 9.3), with manufacturing still only 6 per cent of GDP, about one-quarter of the norm for a country of Saudi Arabia's size and level of per capita income. Meanwhile, at 8 per cent of GDP, agriculture was barely half the size expected and even then required massive subsidies (sanctioned for strategic reasons). The subsidies set wheat prices at 2 to 4 times world levels and led to milk production from sun-shaded cows in the desert around Riyadh.

The Saudi government responded prudently to the oil revenue loss of the mid-1980s by trimming spending to close the budget deficit. However, the deficit was still $10 billion annually in the late-1980s. At first, the government drew on its overseas assets to cover the deficits and smooth the transition to lower oil prices. The infrastructure construction programme was cut back and farm subsidies were reduced. Import tariffs were also raised from 5 per cent to 20 per cent in order to boost tax revenues and to provide an incentive for domestic manufacturers. By the early 1990s the overseas assets which had cushioned against the decline in oil revenues are thought to have fallen by over $100 billion, and Saudi Arabia turned to the IMF for assistance. Although

Table 9.3 *Structure of Saudi Arabian GDP, 1970–87(%)*

	1970	1975	1979	1980	1981	1982	1983	1984	1985	1986	1987
Total GDP (Billion US dollars)	3.89	39.65	74.49	115.17	156.80	154.33	121.06	109.97	93.60	74.86	70.42
Oil sector (%total GDP)	53.74	79.18	55.50	64.81	68.96	64.14	49.74	43.37	37.78	32.61	32.24
Non-oil sector (% total GDP)	46.26	20.82	44.50	35.19	31.04	35.86	50.26	56.62	62.22	67.39	67.76
Share of non-oil output											
Agriculture	12.17	4.83	3.85	3.48	3.50	3.63	4.24	4.57	5.55	6.80	8.02
Manufacturing	5.34	5.56	4.74	4.84	4.57	4.91	5.20	5.81	5.95	5.83	5.91
Construction	11.55	26.84	31.87	32.28	31.67	26.70	23.50	23.73	20.69	18.70	18.44
Other	71.94	57.81	59.54	59.40	60.26	64.76	63.85	65.89	67.81	68.67	67.83
Memo item: Refining and non-oil manufacturing output 1983–7											
Refining (billion US dollars current)					3.908	4.392	2.468	2.474	1.972		
Non-oil manufacturing (billion US dollars current)					3.143	3.649	3.466	2.939	2.819		
Total (billion US dollars current)					7.051	8.041	7.114	5.413	4.791		
Total (billion US dollars constant 1970)					1.190	1.367	1.332	1.277	1.263		

Source: Auty (1990a)

the prolonged decline in per capita income had stabilized, the prospects for renewed economic growth were uncertain in the absence of renewed oil price rises.

Nigerian windfall deployment

In contrast to the overall positive impact of the oil shocks on Saudi Arabia, the Nigerian economy was seriously weakened. Prior to the 1973 oil shock, the Nigerian economy had been growing at more than 8 per cent annually as the country recovered from civil war. Nigeria was, at that time, an agricultural export economy which generated 39 per cent of GDP from farming, based mainly on peasant farms averaging around 2 hectares in size. Nigeria was the leading world producer of groundnuts and palm products in the early 1970s, and the second largest producer of cocoa. The country was also largely self-sufficient in food and was able to feed Sahelian refugees during the severe drought. But manufacturing was not well-developed: it accounted for only 6 per cent of GDP and comprised inefficient import substitution industries, like the assembly of cars and televisions from kits, along with some sugar and tin processing.

During the first oil boom of 1974–8, a relatively large share of the windfall at first went into overseas saving, but the savings were then rapidly run down. During the 1979–81 oil boom, far from saving a fraction of the windfall, Nigeria accumulated overseas debts exceeding $30 billion, which proved too onerous to service when oil prices fell. Investment dominated the absorption of the 1974–8 windfall and went into infrastructure, especially roads and primary education. This investment helped to spread the benefits throughout the entire country and was very popular. Much of the infrastructure spending went on rural roads, while primary school enrolment expanded rapidly from 3.5 million in 1970 to 13 million in 1980. While such investment is sensible, it takes several decades to garner the return.

Domestic expenditure of the oil windfall was already exceeding domestic absorptive capacity when the 1979–81 oil boom occurred. Investment again dominated the windfall expenditure but, in addition to expanding the infrastructure, Nigeria launched an ambitious RBI programme which focused on steel, and the first steps were taken to relocate the federal capital from Lagos to Abuja, closer to the centre of the country (Fig. 9.3). An integrated steel plant was started at Ajaokuta, using domestic coal and iron ore which proved both poor in quality and difficult to transport. A second steel plant was built nearer the main markets in the Delta region, using natural gas to smelt imported ore and scrap. Three steel rolling mills were built to process the steel, but they were scattered throughout the country, for political rather than economic reasons.

These investments did not strengthen the Nigerian economy. The rate of economic growth slowed markedly from 9 per cent before the booms

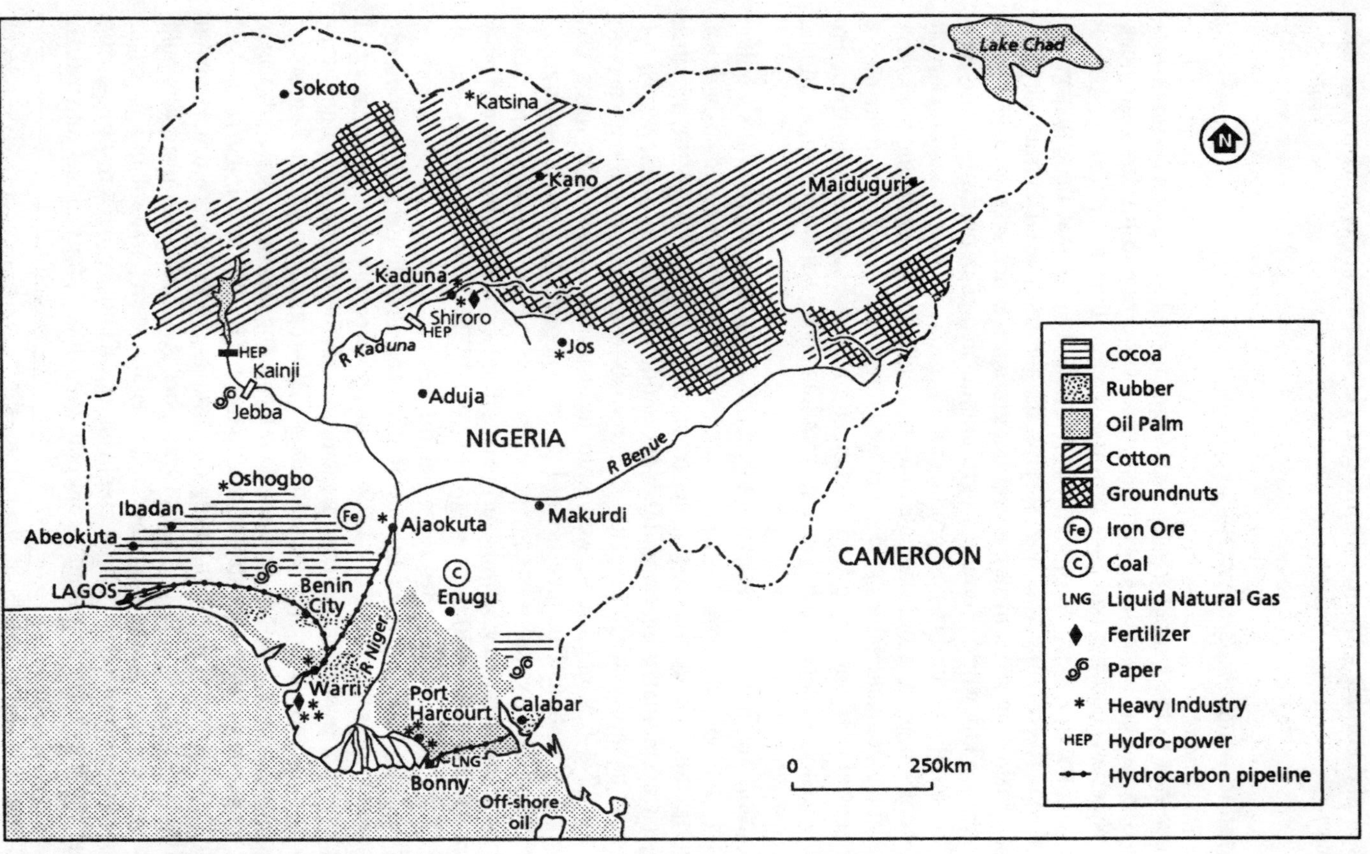

Fig. 9.3 *Nigerian space economy*

(1970–2) to 4 per cent per annum during the first oil boom, and turned negative in the 1980s. By the end of that decade, real per capita incomes had fallen back below their 1970 level in real terms (whereas incomes in resource-deficient Korea and Taiwan had tripled). The manufacturing sector remained weak and required high tariff protection against imports. It generated only 7 per cent of non-mining GDP, about one-third the size to be expected, and mainly assembled imported components. Manufacturing capacity use fell to 25 per cent in the mid-1980s, as oil revenues fell and firms lost both their domestic market and the foreign exchange needed to import inputs (which provided more than 75 per cent of all inputs).

The new RBI plants proved to be white elephants. For example, in the mid-1980s the imported price of steel was $450/tonne, compared with an estimated cost of $1500/tonne at the Delta steel plant, and more than $2000 for the integrated plant at Ajaokuta. The latter had cost $6 billion to build, some four times more than an efficient plant would cost. The state-owned Ajaokuta plant was grossly overmanned several years before steel production began, being largely used as a source of political patronage. Ajaokuta imported both coal and iron ore, because local supplies proved inadequate. It would have been more economical to have abandoned it, but political motives dictated otherwise.

Not only did Nigeria fail to create a competitive manufacturing sector during the oil booms, but Dutch disease effects caused its agriculture to collapse. The size of the farm sector halved to only 20 per cent of GDP in the 1980s, even though it still employed more than half the labour force. Palm oil and groundnut exports slumped, and only cocoa production held up. Worse, food production failed to keep pace with population growth, and young workers left the land for urban jobs in construction and public service. By 1981, food accounted for one-quarter of imports, so that when oil revenues fell, there was insufficient foreign exchange to meet the country's food import needs.

The windfall expenditure on education will take years to produce a return, and may have low efficiency because of the fast rate at which schools were expanded. It is difficult to escape the conclusion that Nigeria might be wealthier and have less debt had the oil boom never occurred. Such a conclusion applies to a majority of the oil-importing countries (Gelb 1988, Auty 1990a), but Chapter 10 shows with reference to Indonesia that this outcome was by no means inevitable.

Overall, however, the gains made by Saudi Arabia among the low absorbers, and by Indonesia (which more than doubled its real per capita income during 1970–90) among the high absorbers, are unusual. The deployment of the oil windfalls was generally disappointing and reflects just one way in which the efforts to reform the LIEO (through commodity price manipulation by a producer country cartel) backfired on the developing countries. Meanwhile, the long-term decline in ore prices triggered by the oil price shocks adversely affected the ore-exporting mineral economies.

Table 9.4 *External shocks, Zambia, Chile and PNG, 1970–87*

Terms of trade	1970–4	1975–9	1980–3	1984–7
Zambia	216.5	101.3	82.2	73.2
PNG	121.2	127.0	93.1	93.4
Chile	192.6	103.4	87.3	80.1

External shocks (% GDP)	Terms of trade, 1974–8	Terms of trade and interest, 1979–83
Zambia	–22.6	–13.6
PNG	5.9	–14.9
Chile	–10.6	–6.0

Source: World Bank (1990c), methodology after Sachs (1985)

Table 9.5 *GDP growth, Zambia, Chile and PNG, 1967–88 (%yr)*

	1967–73	1973–80	1980–8	1973–88
Zambia	2.4	1.2	0.9	1.1
Chile	2.2	2.4	2.2	2.3
PNG	6.5	2.2	2.0	2.1
All developing countries	6.5	5.1	3.9	4.5

Source: World Bank (1990c)

The ore exporters

A common feature of mineral economies is the formulation of over-optimistic projections of mineral revenues. The copper producers misjudged the long-term copper price trend after it rose to an average $1.78/lb during 1965–9 (in constant 1980 prices). Like the oil and bauxite producers, they established a producer group (CIPEC) in an effort to maintain high prices but, in fact, prices fell to an average $1.40 during 1970–4, and then almost halved to 74c during 1975–82 before falling further to 67c during 1983–6, when a recovery commenced. The price decline inflicted large negative shocks on the economies of the copper exporters (Table 9.4).

Table 9.6 *Protection and trade openness in Chile (%)*

Sector			Effective protection			
			1967	1974	1979	
Consumer goods			138.8	189.7	13.2	
Intermediate goods			172.9	139.6	14.0	
Machinery and transport equipment			265.3	96.0	13.0	
Equally weighted arithmetic mean			176.7	151.4	13.6	
Standard deviation			279.0	60.4	1.7	
Memo item: openness						
Year	1929	1951–5	1965–70	1971–3	1974–9	1980–2
Share of foreign trade in GDP	66.3	21.7	24.0	20.3	36.1	32.6

Source: Corbo and de Melo (1987, p. 114)

A comparison of the response of three producers (Chile, Papua New Guinea (PNG) and Zambia) shows the policies which are required for sustainable mineral-driven development. Chile was the most successful of the three copper exporters: it experienced both a strengthening of its economy and a significant fall in its mineral dependence during 1973–90 (Table 9.5). PNG trod water and remained highly mineral dependent. Zambia, like Nigeria (and the other copper producers, Peru and Zaire), experienced a cumulative economic deterioration and higher mineral dependence at the same time as the mines were increasingly running down.

Chile's pragmatic orthodoxy

The election of a minority left-of-centre government in Chile under President Allende in 1970 led to major reforms, which included the nationalization of the copper industry. It also heralded a sharp deterioration in the economy, and when Allende was displaced by the repressive right-wing military government of General Pinochet, bold orthodox macroeconomic policies were adopted. These policies included a depreciation of the exchange rate to restore external balance, and public spending cuts to close the large fiscal deficit. In addition, trade barriers were lowered and prices were freed to change economic incentives and restructure the economy in line with Chile's comparative advantage in resource-based activity. Table 9.6 shows the recovery in the share of trade in the country's GDP through the 1970s. It also traces the dramatic fall in the level of protection of Chilean manufacturing.

But Chilean policy became increasingly doctrinaire during 1978–82, and the government made a serious error which jeopardized economic recovery. Basically, the government wrongly forecast a copper boom beginning in 1979. It failed to set up a mineral stabilization fund (MSF) to mute the domestic absorption of the revenues (as Saudi Arabia had done), and it let the exchange rate appreciate. As tariff barriers on manufacturing were lowered and the exchange rate strengthened, Chilean factories had difficulty competing with imports and got into financial difficulties. When the expected mineral boom failed to materialize, a financial crisis occurred. GDP fell by 14 per cent in 1982 and the government was forced to adopt a more pragmatic orthodox policy, which involved more state intervention.

Economic stabilization required a devaluation of the exchange rate, together with public spending cuts. But in addition, tariffs were temporarily raised for two years in order to boost government revenues and to give the hard-pressed manufacturers a breathing space from foreign competition. The government also intervened in financial markets to avert a banking collapse and set up an MSF in 1985 to smooth the revenue inflows through the mineral cycle of boom and bust.

By the late-1980s Chile had an annual economic growth rate in excess of 5 per cent. It had also reduced its level of mineral dependence, so that mining provided only 45 per cent of exports, compared with 85 per cent in the early-1970s. Manufacturing accounted for one-third of all exports, with large contractions in ISI offset by expansions in RBI like timber products and metals (Gwynne 1985). Agricultural exports also increased substantially, notably the export of out-of-season fruit to northern hemisphere markets. Having controlled Dutch disease, Chile was well-placed to adopt a truly sustainable development strategy.

Undiversified growth in PNG

The experience of Papua New Guinea confirms the importance of pragmatic orthodox macro policy. It also shows that, contrary to the structuralist view, a small country can take on the MNCs and win. But PNG failed to diversify its economy into competitive manufacturing and agriculture.

The prudent macroeconomic policies of the PNG government and its establishment of an MSF helped the country to weather three major economic crises. The first crisis occurred in 1974 when, after the renegotiation of the Bougainville mining agreement in the favour of the PNG government, copper prices suddenly fell. The economic boom associated with the start-up of the large Bougainville mine was abruptly punctured. The second crisis occurred in 1982 when the copper price again collapsed and the MNCs postponed the start-up of PNG's second large copper mine at Ok Tedi. Instead, the MNCs began exploiting the Ok Tedi gold, thereby reducing the chance of the copper deposit being mined in the future. The PNG government forced the MNCs to extract both copper and gold, as originally intended. Finally, the third crisis occurred when Bougainville was abruptly closed in 1989 as a result of a local insurrection.

Although PNG coped well with all three crises, its economy grew slowly and its mineral dependence increased. This reflected flaws in micro policy which restricted foreign investment in plantation agriculture, maintained public spending at twice the normal level for such a country, and set high minimum wages. By the late-1980s PNG wages were 50 per cent above those of Korea, which had five times the per capita GNP. High labour costs combined with a small and scattered domestic market to constrain manufacturing. Meanwhile, forestry operated at one-quarter of its estimated sustained yield, because of management deficiencies, while agricultural growth averaged only 2.3 per cent in the 1980s, one-third of its potential (World Bank 1988). Zambia suffered from the same rent-seeking (high-wage) environment as PNG, but lacked its macroeconomic success.

Zambian regression

The Zambian economy traced a cumulative deterioration because that country was not sufficiently rigorous in adjusting to external and fiscal imbalances which became ever more intractable. Yet Zambia had been cautioned by the UN, upon gaining independence in 1965, over its excessive dependence on copper and of the need to diversify its economy. Unfortunately, as in much of sub-Saharan Africa, policy discriminated against peasant farmers and in favour of state-owned factories and government departments in the cities. The resulting lack of rural incentives combined with the prospect of higher-paid employment city jobs to give Zambia one of the highest rates of urbanization in sub-Saharan Africa.

The 1974 fall in the copper price was equivalent to the loss of 23 per cent of GDP for Zambia (compared with 10 per cent for Chile). The Zambian government devalued the exchange rate and cut public spending (mostly investment), but the non-mining tradables sector was weak and did not respond to large depreciations of the exchange rate. This reflected both the inefficiency of Zambian ISI and the prolonged mismanagement of the peasant farm sector. Although the fertile area on either side of the Rift Valley in Zambia (one-eighth of the land) could meet all domestic agricultural needs (Dumont and Mottin 1983), mismanagement prevented this. For example, the extension of maize throughout much of the country, intended to speed the diffusion of green revolution techniques, displaced crops better suited to local growing conditions (Good 1985). Even with falling copper revenues, the non-mining tradables sector earned less than 6 per cent of the country's exports.

The principal Zambian adjustment to falling copper revenues therefore took the form of foreign borrowing, and the country's external debt tripled to $3 billion during 1974–9. This was equivalent to 100 per cent of Zambian GDP, twice the level for Chile and PNG. When copper prices fell further in 1982, the magnitude of economic adjustment which Zambia needed to make was even greater than in the mid-1970s. The political costs of reform were high, arousing strong opposition from urban-based interest groups, which jeopardized the economic reforms.

Even as Zambian farming and manufacturing remained uncompetitive, mining output contracted from 710000 tonnes to 420000 tonnes between 1969 and 1989 (when copper reserves were estimated sufficient for no more than two more decades). The decline of the state-owned mines resulted from overmanning and excessive taxation to prop up inefficient urban-based activity. The mines had insufficient funds to maintain their equipment. Such corrosion of mining, when it is the only competitive sector, is an alarming feature of mismanaged mineral economies which Zambia shared with Zaire, Bolivia and Peru (Auty 1993).

Sustaining mineral-driven development

Sustainable development of mineral economies calls for:

- replacement of the depleting mineral asset with alternative wealth-generating assets;
- curbs on environmental degradation.

But these goals require the adoption of pragmatic orthodox macroeconomic policies (as followed in Chile after 1982). This is because such policies promote the diversification of the economy away from dependence on minerals by preserving the competitiveness of the non-mining tradables sector. They mute the mineral-driven swings in the exchange rate and thereby limit Dutch disease effects. These policies also encourage MNC investment in mining which uses new resource-efficient and pollution-minimizing technology to help curb environmental degradation.

Replacing the depleting mineral asset

Mineral economies need to set aside an adequate sum out of mineral revenues to replace the depleting mineral asset. This requires that the net revenue from mineral exploitation must first be identified. The net revenue is the total product receipts from mining, less the associated labour and capital costs (Mikesell 1992). Such revenue is then split between an income component, which may be consumed, and a capital component, which must be invested to replace the depleting asset.

The amount set aside as the capital component should be that annual sum which, when invested, will yield sufficient income to replace the mining revenue when ore extraction ends. For example, assuming that the mine has a 20-year life, that the interest rate is 10 per cent and that net mine revenues are $250 000 annually, then the capital component is $37 000 and the income component is $213 000 (Mikesell 1992). This is because $37 000 invested annually will cumulate to $2.13 million at the end of the life of the mine, a sum which yields an income of $213 000 at 10 per cent interest.

In practice, however, the accounting system used is likely to be one which requires a cumulative increase from an initial low level in the figure set aside to replace the depleting assets. This reflects the use of discount rates to compare present and future costs and benefits (see Chapter 12). The higher the discount rate used and the longer the life of the mine, the lower will be the fraction of mineral revenue which must initially be set aside for depletion. It follows that since the mineral reserves of most countries will last decades and a high discount rate is appropriate for developing countries (because the scarcity of capital creates very high returns, or 'opportunity costs'), the share of net revenues required to substitute for the depleting asset is initially very small. This encourages complacency, and also the neglect of any weakness in the non-mining tradable sectors.

Such complacency assumes a predictable smooth transition towards mine exhaustion, whereas in reality a mine can be abruptly marginalized through the discovery of large new low-cost reserves or technological substitution or economic mismanagement. Moreover, mineral revenues tend to fluctuate in a way which, in poorly managed economies, weakens the non-mining tradables, as a result of Dutch disease. This reverses the economic diversification which sustainable development requires and heightens mineral dependence. *In extremis*, as Zambia shows, even the mines may be rendered uncompetitive, so that the mineral resource may become a curse rather than a blessing (Gelb 1988, Auty 1993).

In a severely weakened economy, the sustainable development criteria are subverted by economic crisis management. For example, far from adequately replacing its depleting mineral assets, Zambia set aside insufficient capital to explore for new ore deposits (i.e. to expand its environmental assets), while it also ran down its stock of man-made assets in mining. But Saudi Arabia, PNG and post-1982 Chile show that the Dutch disease effects can be muted through the pursuit of pragmatic orthodox macroeconomic policies, incorporating an MSF which smooths the injection of mineral revenues into the economy.

Environmental accounting can help to distinguish whether growth is sustainable or is based upon environmental asset depletion. A World Bank study (van Tongeren *et al.* 1991) applies environmental accounting to Mexico for the year 1985. Using the World Resources Institute (WRI) accounting system, it shows that a very substantial reduction in the environmental assets occurred within the oil sector in 1985, because the value of depleted oil assets was more than three times the value of oil added to the reserves.

In terms of the national accounts, the starting point for the WRI system is the net domestic product (NDP), which is GDP minus the annual depreciation of man-made capital. It calculates an initial measure of environmental

Table 9.7 *Expenditure distribution with environmental accounting, Mexico 1985*

	NDP		EDP1		EDP2	
	Billion pesos	% Expenditure	Billion pesos	% Expenditure	Billion pesos	% Expenditure
Net product/expenditure	42.061		39.663		36.448	
Final consumption	34.949	83.1	34.949	88.1	34.891	95.1
Capital accumulation, net	4.704	11.2	2.306	5.8	(0.850)	(2.3)
Economic assets	4.704	11.2	4.704	11.9	4.704	12.9
Environmental assets			(2.398)	(6.1)	(5.554)	(15.2)
Exports–imports	2.408	5.7	2.408	6.1	2.408	6.6

Source: van Tongeren *et al.* (1991)

Note: NDP = Net domestic product = GDP – depreciation of man-made capital
EDP1 = NDP – environmental asset depletion
EDP2 = EDP1 – environmental degradation

domestic product, EDP1 (which is the NDP minus environmental asset depletion). This was 94 per cent of NDP in the case of Mexico in 1985. A second measure of environmental domestic product, EDP2, adds a charge for estimated environmental damage from oil exploitation, and was only 87 per cent of NDP.

Turning to the manner in which economic output was absorbed by the Mexican economy in 1985, Table 9.7 shows that the NDP comprised 83 per cent consumption and 11 per cent capital formation (the remaining 6 per cent of NDP was paid abroad to service the country's $100 billion debt). The inclusion of an asset depletion charge under the WRI method changes these figures to 88 per cent for consumption and only 6 per cent for capital formation. Finally, the incorporation of environmental degradation lifts consumption to 96 per cent of EDP and reduces investment to minus 2 per cent (Table 9.7). Yet, in order to secure long-term growth without depleting the capital stock (now adjusted to include environmental capital as well as man-made capital), investment in developing countries should comprise some 20 per cent or more of output.

But a word of caution is required concerning the application of environmental accounting. The results are sensitive to the accounting system used. The deterioration in the Mexican oil assets appears to be less severe using the El Serafy method rather than the WRI system (Ahmad *et al.* 1989). For example, the sum required by Mexico in order to cover oil depletion is only $0.62 per barrel using El Serafy, compared with $4.50/barrel under the WRI system of accounting. Moreover, environmental accounting (like conventional national accounts) omits changes in a third important component of capital, human capital. In a well-managed economy like that of Chile, human capital might well more than amply compensate for depletion of environmental resources, if the proceeds are used for education which leads to improved prospects for technological substitution (Birdsall and Steer 1993). The trade-off between man-made, environmental and human capital assets is discussed further in Chapter 12.

Limiting environmental degradation

Curbing environmental degradation also requires that the cost of environmental damage should be estimated and an appropriate charge made on the mineral revenues. The use of charges to encourage pollution abatement is generally considered to be superior to command-and-control systems, because charges provide incentives to reduce pollution at the lowest cost to society as a whole (Tietenberg 1992).

Auty and Warhurst (1993) show that environmental degradation from mining is an unforeseen legacy of the surge of nationalization in the developing countries (see Chapter 8). Whereas the state-owned mines have tended to under-invest and to make inadequate exploration expenditure, MNCs have

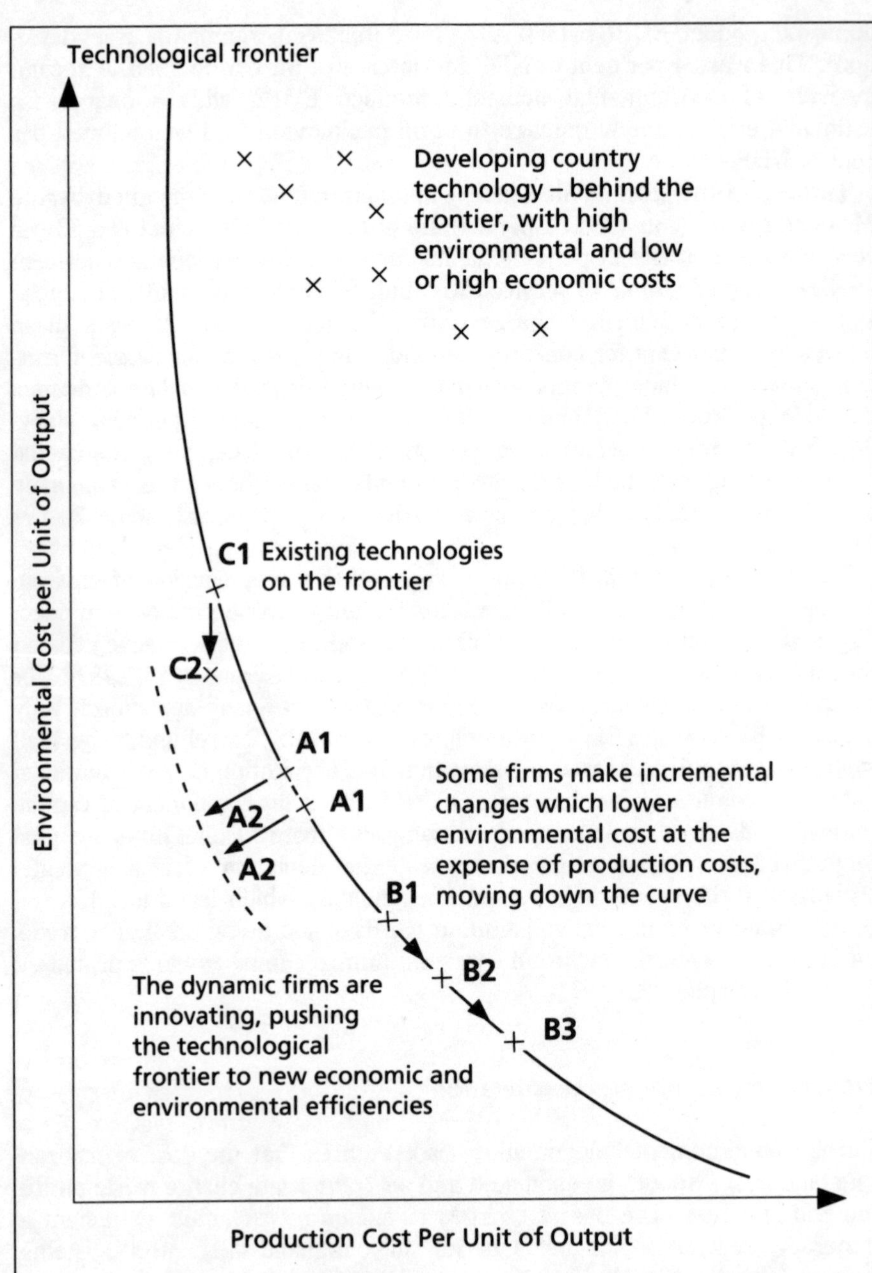

Fig. 9.4 *Environmental trade-offs*

tended to maintain a high level of investment. But MNCs have also experienced growing pressure to adopt environmentally sensitive methods, from shareholders and lending institutions.

Shareholders increasingly call for the 'greening' of MNC investment, a condition which is reinforced by the attachment of environmental conditions to foreign loans from the international lending institutions. In addition, MNCs may adopt high standards of pollution control, in anticipation of even more stringent environmental demands in the future. The early adoption of such standards reduces the risk of possible future litigation over responsibility for environmental clean-ups. It is also usually cheaper to incorporate clean technology in new investments than it is to add (back-fit) such equipment later.

A further advantage in the adoption of clean technology from the outset stems from the fact that it may also improve cost competitiveness. The trade-off between production costs and environmental costs in Fig. 9.4 shows that dynamic firms tend to use new low-cost production methods and pollution control (A_1–A_2). In contrast, many state-owned firms work at lower levels of both economic efficiency and pollution control (B_1–B_2). Moreover, inefficient recovery of ore by state-owned firms has resulted in their mine tailings (waste rock) attracting numerous small firms which extract additional ore using even less environmentally sound methods than those of the state mines. Yet if a strict environmental regime is abruptly imposed, high-cost mines may close and leave the state, rather than the firm, responsible for any subsequent clean-up.

In contrast to most state-owned mining enterprises, MNC subsidiaries tend to adopt their headquarters standard and to press the host government to encourage the widespread domestic adoption of such standards. In Chile, for example, environmental policy initially lapsed because standards were set which were too high to enforce. But as Chile liberalized its politics in the 1980s, the government set up national air quality zones and demanded clean-ups – for MNCs. For example, Exxon was ordered to reduce emissions in 1985, even though this meant that Exxon used stricter standards than those in the USA, while in the same Chilean valley, a state-owned smelter was left to pollute unchecked. Or again: the Chilean state mining firm, Codelco, was ordered in 1986 to set up air quality monitoring and technology to cut sulphur emissions. But the Chilean Treasury, which controlled the finances of Codelco, refused to sanction investment by the state firm in major environmental improvements until the early 1990s. This suggests that the privatization of state mines might lead to rapid reductions in environmental degradation.

Conclusion

The mineral economies present a clear instance of the resource curse thesis. Although the mineral resource provides additional advantages over other developing countries, that advantage has rarely shown up in faster or more equitable economic growth. The mineral resource encourages over-optimistic projections of mineral rents and tax revenues, which lead to lax macro policies. Such policies merely heighten the harmful Dutch disease effects of mineral booms and increase mineral dependence. This reverses economic diversification and retards replacement of the depleting mineral asset, so that one important requirement for sustainable development is violated. The cumulative economic weakness may well be so severe as to override concern for the second criterion of sustainable development, namely, minimal environmental degradation.

The application of environmental accounting suggests that the mineral economies have made inadequate provision for replacing environmental capital, so that conventional investment and economic growth measures overstate the underlying economic performance. But the application of environmental accounting is sensitive to the asset replacement method used, while a case can be made for extending the accounts to include changes in human capital (education levels) as well as man-made capital and environmental capital. It is possible that the depletion of environmental assets to build up human capital may leave the total capital stock available to future generations at a higher level.

Overall, the disappointing results of the mineral economies' efforts to capture greater revenues from their depleting mineral assets, either through producer cartels or the nationalization of state enterprises, provides a salutary lesson concerning the attempts to forge a NIEO. Large transfers of resources from rich to poor countries may easily be squandered and prove counterproductive if absorptive capacity is weak. In contrast, the developing countries of East Asia have sought to turn the LIEO to their advantage with considerable success, as Chapters 10 and 11 show.

PART 6

The role of the state: mid-income East Asia

10

East Asia's development escalator: Malaysia and Indonesia

Whereas economic growth in Latin America, sub-Saharan Africa and most mineral economies was adversely affected by the post-1973 commodity price shocks, many Asian countries, especially those in East Asia, achieved rapid economic growth (Table 1.4). The four East Asian 'dragons' (Hong Kong, Singapore, Taiwan and Korea) industrialized more successfully than Brazil and Mexico and set new best practice for rapid and equitable economic growth. This superior East Asian performance occurred despite the advantages for industrialization which the two Latin American countries held on account of their larger markets and more varied and richer natural resource endowments.

But the success of the East Asian countries was not confined to any specific natural resource endowment. Resource-rich Malaysia, Thailand and Indonesia also achieved rapid economic development (Fig. 10.1). This confirms that Mahon's (1992) speculation on the problems of well-endowed countries and the resource curse thesis (see Chapter 9) lacks universal application. Indeed, Indonesia provides a model of prudent oil windfall deployment. By the end of the 1990s its economy had become sufficiently diversified no longer to warrant the country's classification as a mineral economy.

Nevertheless, this chapter shows with reference to Malaysia and Indonesia that the resource-rich Asian countries also made the error of promoting inefficient industry during the oil booms. However, the overriding priority which their governments gave to pragmatic orthodox macroeconomic management prevented the cumulative deterioration observed in Latin America and sub-Saharan Africa.

A shared characteristic of the successful East Asian countries is an acceptance of pragmatic orthodoxy and of the LIEO (World Bank 1993a). But that acceptance has been combined with astute interventions by some governments to maximize the domestic advantages of international trade. For example, the Korean government intervened substantially to force industrial change, whereas Hong Kong adopted a minimalist position on state intervention, which stressed the provision of infrastructure and left markets to guide investment (Chau 1993).

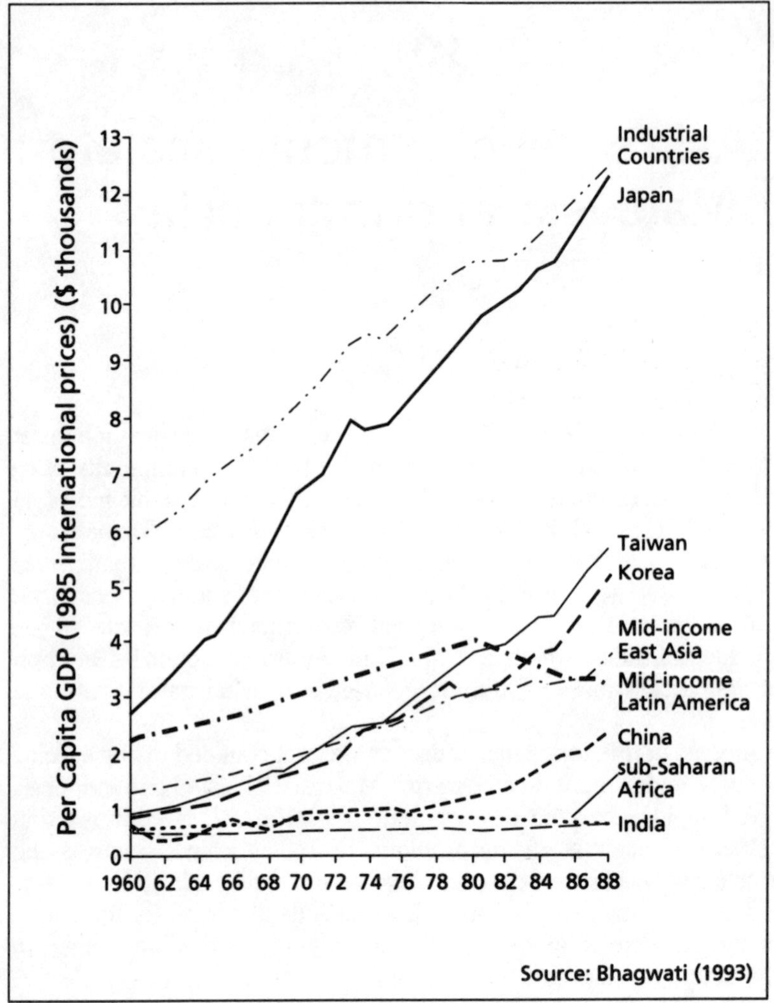

Fig. 10.1 *GDP per capita growth, 1960–88*

The role assigned to MNCs, the *bête noire* of the NIEO, has also varied: Singapore assigned MNCs a strong role within its economy (Soon and Tan 1993), whereas Taiwan and Korea restricted MNCs even more than Brazil or Mexico. Finally, although the East Asian countries have shown a marked preference for private firms, state enterprises have been used to control the 'commanding heights' of the economy, like the steel and petrochemical sectors, and their performance has in some cases been very effective.

There is some evidence of a demonstration effect within the East Asian region. The diffusion of economic development through East Asia presents a

Table 10.1 *The East Asian development escalator*

Country	Area (million sq. km)	Population (million)	Cropland/hd (ha)	Population growth (%/yr)	Population in cities 1991 (%)	GDP growth 1980–91 (%/yr)	Per capita income ($1991)	Literacy (%)	Life expectancy (yrs)
First tier									
Japan	0.378	123.9	0.04	0.5	77	4.2	26 930	100	79
Second tier (NICS)									
Hong Kong	0.001	5.8	0.01	1.2	94	6.9	13 430	100	78
South Korea	0.099	43.8	0.05	1.1	73	9.6	6330	96	70
Singapore	0.001	2.8	0.00	1.7	100	6.6	14 200	100	74
Taiwan	0.036	20.4	0.02	1.4	n.a.	6.5	7600	44	n.a.
Third tier (mid-income)									
Malaysia	0.330	18.2	0.27	2.6	44	5.7	2520	78	71
Philippines	0.300	62.9	0.13	2.4	43	1.1	730	90	65
Thailand	0.513	57.2	0.40	1.9	23	7.9	1570	93	69
Fourth tier (low-income)									
China	9.561	1150.0	0.08	1.5	60	9.4	370	73	69
Indonesia	1.905	181.3	0.12	1.8	31	5.6	610	77	60
Fifth tier (laggards)									
Cambodia	0.177	8.3	0.37	2.5	12	n.a.	n.a.	35	51
Laos	0.247	4.3	0.22	2.7	19	n.a.	220	n.a.	50
Myanmar	0.657	41.7	0.24	2.1	25	2.3	n.a.	81	63
Vietnam	0.325	66.7	0.10	2.2	22	n.a.	220	88	64

Sources: World Bank (1993b) except: Taiwan = CEPD (1991)
Cropland/ha = WRI (1992)
Fifth tier = WRI (1992)

cascade (sometimes referred to as the 'flying geese' formation). It runs from Japan, through the four so-called dragons (Korea, Taiwan, Singapore and Hong Kong), to a third tier comprising the Philippines, Thailand and Malaysia. Indonesia and, more recently China, form a fourth tier, with a residual group, including Vietnam and Myanmar, yet to open their economies to the economic advantages of international trade (Table 10.1).

The cascade metaphor can be replaced by one of a sequence of escalators which reflect the changing comparative advantage of sub-sectors of manufacturing in each country. Rapid economic progress requires the correct identification of the appropriate escalator for a specific stage of development. Unlike countries pursuing an autarkic policy, East Asian countries have tried to avoid alighting on a specific sectoral escalator too early. They have also sought to avoid retaining declining sectors for too long. Rather, East Asian countries are aware of their position in the development cascade. As per capita incomes rise and competitiveness shifts away from labour-intensive sectors, they anticipate which activity they may expect to capture from those countries in the tier above, and which activity they must cede to the tier below.

This chapter and the next explore the reasons for the success of the East Asian countries. The present chapter focuses on the ability of Malaysia and Indonesia to make good use of their natural resource endowment to diversify. They did so, first, into a range of primary product exports (as sub-Saharan Africa needs to do) and then into competitive manufacturing (as is required of Latin America and South Asia).

This sequence confounds the pessimism of the structuralists, so that it is appropriate to begin this chapter with a discussion of two aspects of international trade. The first is an analysis of why the more outward-oriented economies of East Asia coped more successfully with the debt crisis than the Latin American countries. The second aspect is export base theory, which analyses the underlying process of economic diversification based on primary product exports. The case studies of successful diversification by Malaysia and Indonesia then follow.

Differing responses to the 1982 debt crisis

The debt crisis which broke in 1982 is associated with Latin America and sub-Saharan Africa, rather than with Asia. By 1982, the Latin America countries had accumulated overseas debt in excess of $350 billion, of which the three largest countries, Brazil, Mexico and Argentina, accounted for $104 billion, $100 billion and $46 billion, respectively. The combined external debt of the countries of sub-Saharan Africa was $200 billion. Yet the absolute size of the debt is less important than its size relative to the economy as a whole. Table 10.2 shows that, when expressed as a percentage of GDP, the Latin American debt was not much larger than of the East Asian countries which successfully

Table 10.2 *Comparison of East Asian and Latin American responses to 1979–83 shocks*

	% change in GDP Average 1970–81	Average 1981–94	External economic shocks[a]	Debt/GDP ratio 1981	Taxes as % of GDP 1982	Government spending in GDP 1982	Export as % of GDP 1965	1983	Debt service ratio[b]	Urban population ratio 1980
South Korea	8.1	7.8	–3.8	27.6	19.1	19.5	9	37	90.1	55
Malaysia	7.9	6.7	4.8	27.8	29.2	41.0	44	54	16.9	29
Thailand	7.1	5.7	–3.3	25.7	13.9	19.9	18	22	58.1	14
Indonesia	8.1	4.8	6.2	24.1	22.2	23.5	5	25	na	20
Brazil	7.7	1.2	–5.0	26.1	26.1	21.8	8	8	132.6	68
Argentina	1.7	0.0	1.6	31.6	16.5	21.6	8	13	214.9	82
Mexico	6.8	–0.9	1.2	30.9	17.0	31.7	9	20	161.8	67
Venezuela	3.5	–2.0	16.2	42.1	29.3	29.6	31	26	117.8	83
Peru	3.4	–2.0	–4.2	44.7	16.8	18.0	16	21	122.2	67
Chile	3.7	–2.1	–6.2	47.6	32.0	37.6	14	24	153.3	80
Weighted averages										
East Asia	7.8	5.8	1.1	25.9	20.6	23.7	13	32	61.7	32
Latin America	5.6	–0.4	–0.1	31.3	22.2	25.8	11	15	153.8	72

Source: Brookings Institution

[a]Average annual change in income as a % of GDP comparing 1976–8 with 1979–83

[b]Average 1980–3

rode out the debt crisis. For example, whereas Mexican foreign debt was equivalent to 31 per cent of GDP, that of Korea and Malaysia was 28 per cent of GDP.

But when the debt is expressed in relation to exports, in order to provide a measure of the capacity of the economy to earn foreign exchange to service the debt, then Latin American debt appears much more onerous. The average debt/export ratio for Latin America was 272 per cent, compared with 52 per cent for East Asia. As a result, whereas East Asia (excluding the Philippines) was able to service its debt and maintain economic growth, most Latin American countries (with the notable exception of Colombia), as well as those in sub-Saharan Africa, experienced a decline in their per capita incomes during the 1980s in what became known as the 'lost decade'. The East Asian economies averaged 6.2 per cent annual per capita income growth during the 1980s (World Bank 1992a), compared with an annual contraction of 1.1 per cent in the Latin American region (IADB 1992).

As Chapter 2 noted with reference to sub-Saharan Africa, the cause of the debt crisis is disputed. Some blame a deterioration in the external economic environment, while others single out flawed domestic policies. Those who regard external factors as responsible point to the fact that the second oil shock reduced the capacity of the Latin American countries to import. Yet Table 10.2 shows that the East Asian countries experienced negative trade shocks which were of a similar size to those which hit Latin America; moreover, the oil shocks were positive for Latin American oil-exporters like Mexico and Venezuela, but they too ran into debt service problems.

A second external factor cited as a cause of the debt crisis, the unexpectedly sharp rise in world interest rates between 1980 and 1982, meets with a similar rebuttal. Table 10.2 shows that although higher interest rates were equivalent to the loss of 1.3 per cent of GDP in transfers abroad for the Latin American countries, the figure for East Asia was 1 per cent, a similar order of magnitude. (Actual transfers abroad could be higher, however, where a collapse in confidence required sizeable repayments of principal in addition to interest.)

Advocates of the flawed domestic policy thesis, like Sachs (1985), point out that foreign borrowing by developing countries is sensible only if the capital is prudently invested. This means that loans should be invested in either traded goods, such as agriculture and industry, which can generate foreign exchange to service the debt, or in infrastructure which promotes improved productivity in the tradable sectors. Some advocates of the flawed policy thesis argue that the Latin American debt crisis was caused by excessive state intervention. But this view is not supported by the data in Table 10.2, which show that the share of government expenditure in the GDP of Latin American countries was 26 per cent, similar in size to that of East Asian governments.

It is not the scale of state intervention which is critical, so much as its quality. Latin America invested its foreign loans less efficiently than East Asia, with some loans finding their way via state enterprises into current spending to finance public sector deficits (to avoid raising taxes). Latin America was not able to earn sufficient foreign exchange to service the debt because of the slow maturation of the investments. Autarkic industrial policies created large uncompetitive industrial sectors and drove trade down to a relatively small fraction of GDP in the Latin American countries. For example, whereas the share of exports in Latin American GDP rose from 11 per cent to only 15 per cent between 1965 and 1983, in the case of East Asia the share rose from 13 per cent to 33 per cent. For Korea, the export share rose from 9 per cent to 37 per cent over the same period, allowing the country to grow out of its debt, as shown in Chapter 11.

It is difficult to escape the conclusion that the cause of the Latin American debt crisis, like that of sub-Saharan Africa, is rooted in domestic policy errors (Sachs 1985). But before examining the policies which sustained diversified economic growth in Malaysia and Indonesia, a review of export base theory will clarify the relationship between primary product exports and economic growth. It will help to explain why the resource-rich East Asian countries succeeded with primary product export-led growth, whereas their counterparts in Latin America and sub-Saharan Africa did not. The discussion shows how governments can participate in global trade and escape the dependent syndrome feared by dependency theorists while also avoiding recourse to the riskier interventionist policies advocated by the reformist structuralists like Prebisch.

Export base theory and economic growth

Export base theory (or the staple theory, as it is also known) was developed in the context of the 'unsettled' regions of North America by Innes (1920), North (1955) and Watkins (1963). It was formulated in order to explain the growth of diversified prosperous regional economies based upon the export of primary products, rather than upon industrialization. The export base model can be described in terms of linkages, following Hirschman (1977). Four categories of linkage are triggered by the expansion of exports of a primary product. They are:

- backward linkage (the establishment of firms to provide inputs to the export staple);
- forward linkage (the establishment of firms to process the staple product prior to its export);
- fiscal linkage (the spending of government tax revenues levied on the staple);
- final demand linkage (the activities set up in response to the local spending of wages and profits by labour and owners).

Export base theory sees development as occurring in a five-stage sequence which begins when a mining company, or plantation or group of yeoman farmers identifies a product for which the region has a comparative advantage and begins to export it. The second stage is characterized by expanding production, which yields external economies (such as improved shipping facilities) that lower the average cost of production and further boost the region's competitiveness. In this second stage, investment occurs mostly in the export product, but in the third stage additional productive linkages are triggered by the demand for packaging, machinery and spare parts (backward links) and for additional processing (forward links).

In the fourth stage, the fiscal and final demand linkages become more important, as capital overflows strongly into the non-basic or residentiary sector of the economy. Examples include local brewing and furniture production, some of which may eventually be exported. Meanwhile, a growing share of government spending occurs on basic infrastructure and, perhaps, on efforts to speed economic diversification. Maturity is reached when the export product no longer dominates the regional economy: the economy has become sufficiently diversified into services plus a range of other primary products and manufacturing, so that the original export staple has lost its economic dominance.

Baldwin's region: South and Midwest

Historically, the successful diversification of regions producing primary products has worked in temperate regions (like the Midwest, the Pacific North-West, the Pampas, south-eastern Australia and the Witwatersrand region of South Africa) but not so well in the tropics. This puzzle was analysed with reference to the United States by the economist, Baldwin (1956). Basically, Baldwin sought to explain the greater dynamism of the Midwest compared with the South. He concluded that contrasts in the production function (ratio of capital to labour) of the principal staple of each region (cotton in the South and grain in the Midwest) was the key difference.

The plantation economy of the South required much initial capital to establish production, so that the barriers to entry were high. It used mostly unskilled labour thereafter, because there was little opportunity to substitute capital for labour over time. This rendered the production function of the subtropical South inflexible. In contrast the grain crops of the temperate Midwest region required a small initial capital investment (so that entry was easy) and were highly responsive to the application of small increments in investment over time. This gave the Midwest region greater flexibility than the South. The different patterns of linkage triggered by the expansion of exports of each staple explain the superior economic performance of the Midwest.

In the case of the South, there are no alternatives to subsistence for labourers, given the high barriers to entry in many tropical crops. Labour is therefore drawn from the subsistence sector into the plantation sector at a low wage. Capital for the plantation sector is supplied by a wealthy elite. The resulting social structure is therefore one in which the distribution of income is highly skewed: there are a few very rich individuals and many poor people.

The unequal income distribution stunts fiscal and final demand linkage for two reasons. First, the elite oppose taxation because it would be levied almost exclusively on them, given the nature of income distribution. Fiscal linkage is therefore stifled. Second, the elite can afford to import foreign goods whereas the masses have little surplus for spending. As a result, few opportunities are created for local firms to set up in order to supply manufactured goods and final demand linkage is low.

The spatial pattern which such an economy creates is one designed to ship out the staple, rather than to promote local exchange. There are few centres for local trade, so that there is no central place hierarchy and the tiny domestic market is fragmented among the largely self-sufficient plantations. This settlement pattern (and social system) is unconducive to the emergence of regional entrepreneurs who might flourish and help diversify the economy. Rather, profits from the staple are either ploughed back into the export staple, or expended abroad on luxury imported goods or on foreign investments (which may eventually rival the regional economy from which the investment derives). This historical model shares common characteristics with dependency theory.

In contrast, the low barriers to entry of temperate grain production cause the Midwest region to be characterized by many small producers. In addition, small increases in investment can be made which steadily raise crop yields, productivity and incomes. These rising incomes are, moreover, relatively evenly distributed across the entire population. Such a community supports taxation to improve the infrastructure, including local transport and education, because all can expect to benefit from it. The pattern of settlement which develops is Christalleran (see Chapter 7) and creates a growing domestic market, first for basic consumer goods and then for a widening range of increasingly sophisticated producer goods such as farm machinery. The more egalitarian Midwest society also provides a stream of entrepreneurs with the capital to exploit the proliferating economic opportunities.

Neglected aspects of the export base model

If Baldwin's model is applied at a global level, under the assumption that the South reflects tropical staples and the Midwest temperate ones, the disappointing diversification of regions dominated by capital-intensive staples such as cotton, palm oil, sugar cane and minerals can be explained. But the export base model has more difficulty with other tropical products, such as coffee and cocoa, where the barriers to entry are lower and conditions more closely approximate those of the Midwest staple. This implies that the Baldwin model requires modification. Two key changes are: adding the political economy, and noting the ease with which staples can be substituted.

Taking note of the political economy, Lewis (1978) has argued, for example, that the elite in the US South and other tropical regions came to rely heavily on cheap labour. They first used slave labour to establish the plantation system, and when slavery was abolished in the nineteenth century they mobilized the labour surplus of South and East Asia. Lewis argues that in order to attract Asian labour, the nineteenth-century plantation owners needed to provide an income just above that earned by the Asian farmer.

Similarly, the Midwest settlers also looked for an income greater than their alternative opportunity. In their case, it was an income which was higher than that which they could earn on the more productive farms of Europe. European farms were more productive than Asian farms because of the earlier onset of the application of scientific farming techniques (the green revolution). Lewis estimates that European farms yielded around 4000 pounds of grain per hectare in the nineteenth century, compared with only 1700 pounds of grain on the near-subsistent Asian farms.

Lewis (1978) notes that during the nineteenth century there were two large global migration streams. One stream was white and increasingly expensive, and flowed from Europe to the temperate lands. The second labour stream was coloured and cheap, and flowed within the tropical regions. It was drawn first from Africa (under slavery) and then from Asia (under the indentured labour system). Lewis also notes that white workers were well aware of the threat to their higher wages posed by any mixing of the two streams, and they applied strong political pressure to ensure that the two labour streams did not intertwine.

The agricultural revolution in the temperate lands exerted pressure for increased incomes for (white) farmers in those areas. Until the mid-twentieth century, when a green revolution in tropical food grain crops began, no such wage pressure occurred in the tropics. This rendered labour relatively cheap in the tropics. This cheap labour reduced the incentive of plantation owners to substitute capital for labour. One important implication of this is that, contrary to Baldwin, the production function of most tropical crops does not inevitably depress worker incomes. For example, when workers were unionized in the British West Indies in the early-1940s, the productivity of the sugar plantations increased substantially in order to accommodate rising wages (Auty 1976). It was not the capital intensity of the production function which depressed wages in the tropics, but the lack of an incentive by tropical capitalists to substitute capital for cheap labour.

The second modification to Baldwin's model involves allowing for the flexibility of crop substitution. It is the flexibility of crop substitution as opposed to the flexibility of the production function which differentiates between tropical and temperate staples. The Midwest (temperate) staple could be more easily changed in the face of unfavourable prices than many tropical products. This is because the leading temperate crops are perennial and require little processing before export. This means that temperate crops not only incur much lower investment costs, as Baldwin notes, but they can also be more easily switched.

If the prices of wheat declined in the Midwest, farmers there could turn to maize, beef or milk. However, if the price of tropical crops fell, the fixed investment in the cotton gin, sugar cane mill, palm oil plant or coffee bushes would create inertia. Such inertia would encourage efforts to cut costs (in line with the Prebisch thesis), rather than to substitute a more profitable staple. For example, nineteenth-century West Indian plantation owners stifled peasant agriculture because of fear of competition from that sector for land, labour and government assistance (Adamson 1972). They sought to ride out depressed conditions by cutting costs and *discouraging* diversification. This increased the likelihood that tropical regions would become locked in a staple trap, experiencing slow economic growth or even collapse.

Summarizing, Baldwin's model provides valuable insights into the causes of the disappointing economic performance of tropical regions, but he overemphasizes the role of the production function. The production function of tropical staples was not inherently rigid. The rigidity of tropical staples stemmed more from low incentives for raising labour productivity and the difficulty of changing the staple. Low wages in the tropics, by limiting domestic purchasing power, also reduced the scope for diversifying into manufactured goods, as Mellor (1976) noted.

Prospects for primary product export-led growth

If the preceding analysis is correct, the disappointing primary export-led growth in the tropics has its roots in specific historical circumstances: it is not inevitable. The rising productivity of tropical food grains production associated with the diffusion of the green revolution can boost output and raise incomes substantially (see Chapter 4). Such productivity trends, when combined with a declining rate of population growth and appropriate macro policies, will remove the drag of cheap farm labour on tropical wages and push such economies towards the labour market turning-point.

These improved prospects also hold for another important and historically disappointing tropical staple, capital-intensive mining. This is because the attainment of political independence has reduced the outflow of revenue from mining regions to service foreign capital, which occurred during colonial times. Local ownership has increased. The example of Papua New Guinea shows that, even in small countries, independent governments can purchase part of the equity from MNCs (a prudent 20 per cent stake, in the case of PNG) and also tax away the economic rent for infrastructure improvements. The main problems for tropical mines now are the reluctance of MNCs to invest in mismanaged economies, and the difficulty which many governments display in deploying mineral taxes effectively.

Malaysia: successful economic diversification

Primary sector diversification

Plantations have played an important role in Malaysian economic diversification. Many African and Asian countries have retained plantations and sought to expand them, whereas the emphasis in the Caribbean and Central America has been upon securing their abolition or reform. The explanation for this divergence lies in the fact that in the first two regions the plantation is recent, and has few associations with slavery. In fact, the modern plantation has proved to be a flexible institution, with many benefits for developing countries (Graham and Floering 1984).

New plantations have been established where land has been in plentiful supply, in South-East Asia and sub-Saharan Africa. The plantation brings capital for investment in transport to open up an area to trade, and also in the machinery needed to process the crop for global markets. Plantations also require skilled labour, which is increasingly likely to involve the training of local people in farming, engineering, chemistry and management skills. Additional benefits stem from the fact that the exported product earns foreign exchange to buy vital imports of machinery and oil. Export taxes, income taxes and corporation taxes provide extra government revenues.

Graham and Floering (1984) stress the importance of the nucleus plantation for developing countries. This concept implies a division of functions between the central plantation and the peasant farmer. Those activities which benefit from economies of scale, such as plant breeding, processing, and exporting, are handled by the plantation, which also farms a relatively small area of land itself (the nucleus). That area serves as a model for surrounding peasant farmers, who copy appropriate productive practices and send their crops for processing and marketing to the nucleus plantation. In this way, modern skills and methods diffuse through the economy without alienating large tracts of land to 'outsiders', whether from overseas or from elsewhere within the country.

Under British rule, Malaysian economic growth was narrowly based on rubber plantations and tin mines. The rubber plantations were initially owned by companies based in north-west Europe. They attracted large numbers of Indian workers and also remitted substantial profits from Malaysia until the 1940s (Thoburn 1977). But Malaysia subsequently pursued a policy of repatriating ownership, buying out the overseas companies and increasing domestic revenue retention in the process.

Meanwhile, the linkages from the tin mines were more beneficial to the local economy than was normally the case for tropical mines. This is because the tin mines tended to be relatively small-scale operations with a large labour force (some 200 000 mainly Chinese miners in 1900, who produced 50 000 tonnes of tin, more than half the world's production). Ownership was largely

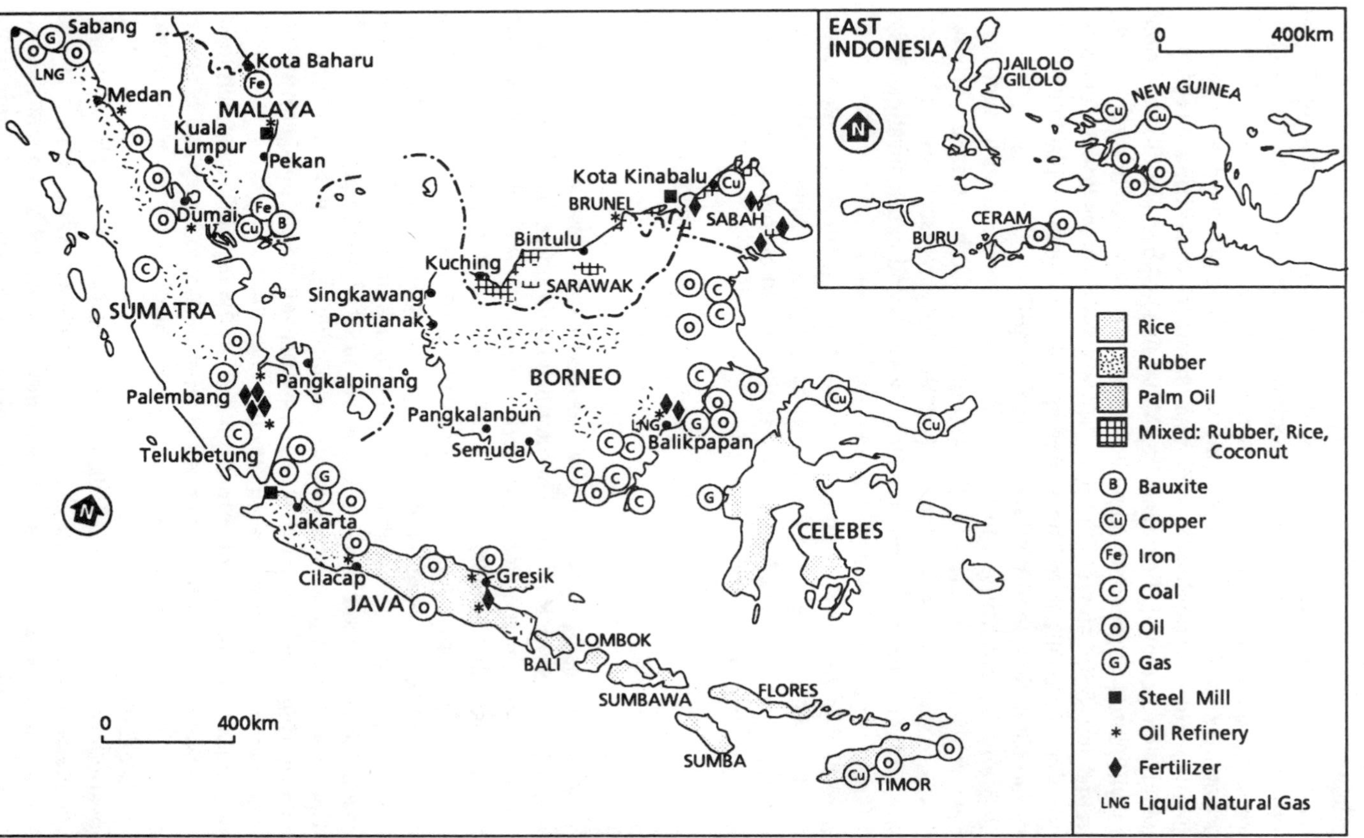

Fig. 10.2 *Economic activity in Malaysia and Indonesia*

domestic and the sector exhibited high backward linkages for machinery and other inputs.

Unlike many sub-Saharan African countries, Malaysia has steadily broadened the range of its primary product exports. As the analysis of export base theory suggests, this has brought advantages in terms of increased flexibility for coping with long-term price changes. As rubber prices fell with postwar competition from synthetics, so Malaysia was able to shift into crops with higher growth potential. For example, Malaysian rubber plantations began diversifying into fast-growing palm oil, for which the country has very strong natural advantages. Later still, cocoa offered a third diversification opportunity so that, in contrast to the stereotype of a monocrop plantation, the Malaysian plantation often produces several crops.

The Malaysian government fostered plantation diversification by undertaking research into crop varieties. It also encouraged other primary product exports, such as timber in Borneo (subsequently replaced with plywood exports, to boost domestic added value), as well as oil and natural gas (Fig. 10.2). The government has also expanded land settlement for rice production, but here its motives were paternalistic rather than commercial. This reflects the fact that the rice sector is dominated by indigenous Malays who, in the face of the more dynamic Chinese community, risked becoming second-class citizens in their own country.

Industrial diversification

At the same time as Malaysia's primary product export base was being diversified (reducing the risk of the economy becoming enmeshed in the 'staple trap' of excessive dependence on a single commodity, with all the boom and bust risks which that entails), the government encouraged industrialization. Industrial policy followed the East Asian model, with an initial emphasis on import substitution through the 1960s quickly giving way to competitive exports of labour-intensive textiles and electronics in the 1970s. As a result, the manufacturing sector increased its share of GDP from 8 per cent in 1960 to 24 per cent by 1985. Meanwhile, per capita income grew by 4.8 per cent annually during 1965–86, so that Malaysia became more prosperous than countries such as South Africa and Mexico.

But as oil revenues expanded through the late-1970s, Malaysia determined to deploy part of the windfall revenues in a big push into heavy and chemical industry (HCI). Key HCI projects included petrochemicals based on natural gas and oil, a gas-based steel plant, as well as an automobile plant and associated engineering works in the Klang Valley, west of Kuala Lumpur. The state firm Hicom executed these projects; like CVG in Venezuela (Chapter 7), it proved inadequate and imposed strains on the domestic economy in the mid-1980s.

As with HCI big pushes elsewhere, inflation accelerated and the current account deteriorated and a sharp deflationary contraction in the economy was

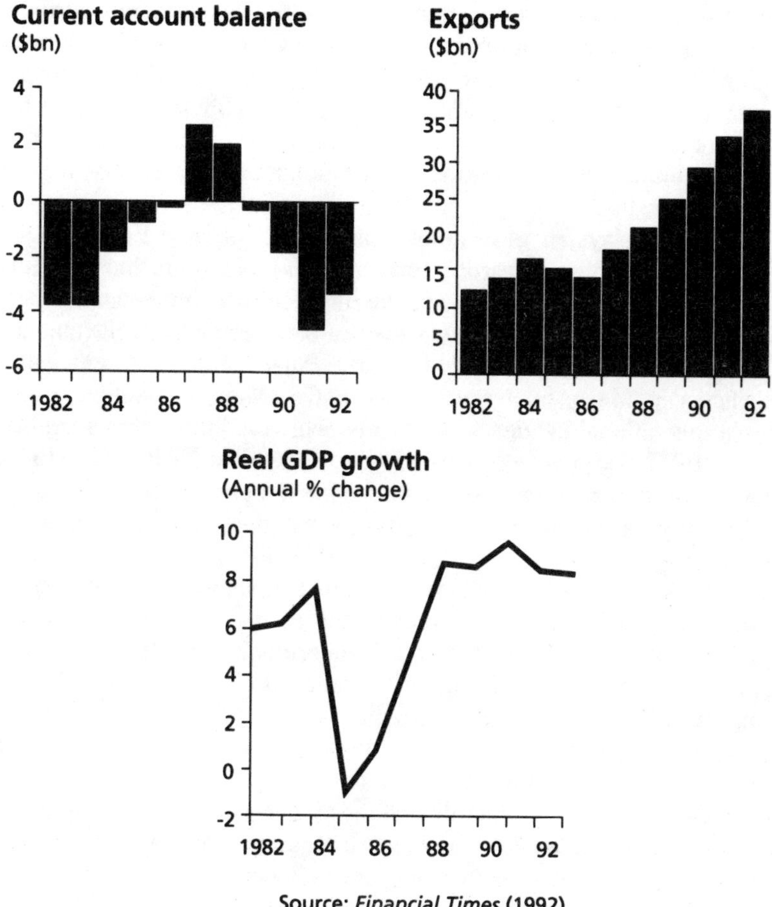

Fig. 10.3 *Malaysian economic performance, 1982–92*

required (Fig. 10.3). For a while, the big push threatened to deflect Malaysia from its exemplary economic performance (Auty 1990a). The Malaysian ICOR deteriorated to 7.2 during 1981–5, so that investment efficiency was half that of the previous decade. The expanded public sector was especially inefficient (Salleh and Meyanathan 1993). However, appropriate macroeconomic measures were promptly applied and the government scaled back its own investment. The mid-1980s' stabilization policy was successful, and Malaysia resumed the role of an emerging second-generation NIC, following the four dragons.

The Malaysian achievement is the more admirable because behind the successful economic growth, there have been tensions between the principal ethnic groups. As noted earlier, one legacy of the Asian labour surplus has been the migration of indentured workers from China and India. Almost two-fifths of Malaysia's population is Chinese, and another twelfth is Indian. The domination of commerce by these two groups (notably the Chinese) threatened the native Malays. The Chinese, in turn, have been apprehensive of reverse discrimination, a fact which caused Singapore to break away from the Malaysian federation.

In 1968 the Malaysian government introduced its New Economic Policy (NEP) in order to enhance participation by native Malays in the economy via state enterprises and other forms of state intervention. Chinese businessmen, in particular, resented the NEP because of its potentially inefficient use of resources by state enterprises, which have proved much less efficient than private firms in Malaysia. Nevertheless, native Malays benefited from the decline in the national incidence of poverty, which fell from 46 per cent of the population to 17 per cent between 1976 and 1990. The NEP reduced income inequality (the gini coefficient fell from 0.53 to 0.43 between 1975 and 1989), but the per capita income of native Malays was still only 82 per cent of the national average.

However, the economic slowdown in the mid-1980s which accompanied the over-ambitious HCI big push exacerbated ethnic tensions. It elicited a pragmatic response from the Malaysian government, which scaled down its commitment to the NEP and sharply reversed the trend towards increased state intervention. Such pragmatism lies at the root of Malaysia's rapid economic transformation from a heavy dependence on foreign-owned rubber plantations into a rapidly industrializing country.

The Malaysian economy has diversified first around a widening range of primary product exports and then through manufacturing. Diversification, by widening the range of competitive sectors, enhances flexibility and improves the capacity of an economy to adjust to sudden shocks, thereby maintaining a rapid rate of economic growth. A similar pattern can be seen in Indonesia on the next rung down of the Asian development escalator (Table 10.1).

The oil cycle in Indonesia, 1970–90

After independence, the Indonesian government stressed political unity over economic development, until a military coup in 1966 installed General Soeharto as leader with a strong commitment to economic growth. Indonesia had many advantages as a populous country with a resilient agricultural sector and substantial oil reserves. As economic recovery got under way, investment was a respectable 22 per cent of GDP during 1967–72, and annual GDP growth exceeded 8 per cent. The economy was relatively diversified, drawing only 31 per cent of its exports from oil, compared with 54 per cent in the case of Nigeria, another populous oil-rich country (Gelb 1988).

The Indonesian economy experienced three external shocks (in 1973, 1979 and 1986) which were each of a similar magnitude, at around 15 to 20 per cent of GDP per annum. The 1973 and 1979 oil shocks were positive, whereas the 1986 shock was negative and built up to 15 per cent of GDP in 1986–8 from a level of 3 per cent during 1982–5. Although the shocks were similar in size to those experienced by Nigeria, Nigerian per capita GDP declined sharply (as Chapter 8 shows), whereas Indonesia almost doubled its per capita income.

Unlike Nigeria, Indonesia benefited from the continuity of a political regime which was committed to macroeconomic orthodoxy and the pursuit of prudent fiscal and exchange rate policies. The priority accorded to macro performance limited the damage done by rent-seeking groups. Interestingly, the Indonesian government did intervene strongly and inefficiently in the manufacturing sector during the oil booms – clear evidence of the resource curse thesis – significantly raising both protection and scope for rent-seeking. But whenever fiscal or trade imbalances threatened the country's long-term growth, macroeconomic stabilization was accorded the first priority (Pinto 1987, Gillis 1984), even if that meant withdrawing the benefits which vested interests derived from rent-seeking behaviour.

The oil booms, 1974–82

Indonesia prudently saved one-third of its first oil windfall and its domestic absorption was further muted by a scandal which engulfed Pertamina, the state oil corporation. Run as a fiefdom of one of the more powerful generals, Pertamina became a sprawling conglomerate which covertly accumulated $10 billion of foreign debt that required servicing. This slowed the rate of domestic absorption of the 1974–8 oil windfall and delayed many Indonesian RBI projects until the second boom. That delay was probably beneficial, given the performance of the state-dominated RBI projects which did proceed (Auty 1990a).

During the 1974–8 oil boom, Indonesia increased the share of consumption in GDP only slightly. Public investment rose to offset a decline in private investment, so that the rate of capital formation increased over the pre-shock rate by around one-fifth. One-quarter of Indonesian public investment during the first oil boom went on infrastructure, including rural projects. The rural areas received investment, because the government was mindful of the fact that rural riots had been instrumental in the downfall of its predecessor. Indonesia raised farm output and productivity by investing in irrigation and green revolution techniques, which transformed the country's food deficit into self-sufficiency.

Two-fifths of the development investment was allocated to the hydrocarbon sector to prolong oil production and diversify both foreign exchange earnings and tax revenues through liquified natural gas (LNG) exports. Two large LNG projects were built in northern Sumatra and eastern Kalimantan, helping to expand and diversify the country's hydrocarbon markets. About one-sixth of the development investment went into metal production (steel and aluminium) and a similar sum was invested in other industries to broaden the manufacturing sector.

Prudent windfall deployment at first characterized the second oil boom, when more than 40 per cent of the windfall was saved abroad. But, like Malaysia, Indonesia found it difficult to resist the pressures for over-rapid absorption. Private consumption, public consumption and investment all increased sharply: inflation accelerated and the exchange rate appreciated. Dutch disease threatened the competitiveness of the erstwhile efficient food grain sector and further delayed infant industry maturation.

Although Indonesian manufacturing grew rapidly during the oil booms, it mistakenly emphasized capital-intensive RBI rather than labour-intensive light industry and was far less efficient than it could have been (Hill 1990). During the first boom, as in Nigeria, the exchange rate appreciation led manufacturers to use the threat of Dutch disease to justify extending protection – and rent-seeking opportunities. The delayed RBI projects were also sanctioned at the start of the second boom, only to be curtailed when oil prices weakened. The RBI projects which did proceed experienced mixed fortunes, with fertilizer generally superior to steel (Auty 1990a).

The protection afforded to Indonesian manufacturing spawned a dualistic sector, in which some firms used the rents to postpone maturation and remain inefficient. Others became efficient and (thanks to protection) highly profitable (Flatters and Jenkins 1986). The large state-owned manufacturing sector was relatively inefficient (Hill 1982).

Post-boom adjustment: reducing oil dependence

The long-term commitment of the Indonesian government to macro orthodoxy led to timely shifts which corrected the fiscal and current account imbalances when the booms waned. It also prevented Indonesia from reaching the extremes of oil dependence seen in Nigeria, and eventually led to a sharp reduction in the importance of oil by the 1990s. Having devalued in 1979 and 1983, Indonesia made a third large devaluation (of 60 per cent from its 1983 level) after oil prices fell in 1986–7. These multiple adjustments contrast to the single massive devaluation of the Nigerian currency to one-quarter of its 1984 level.

The Indonesian government also made prompt cuts in public spending and broadened the tax base during 1984–6 to boost the non-oil tax share. It drew upon foreign loans to smooth the adjustment after 1982 and again after 1986. This pushed the total debt towards 50 per cent of GDP, but a severe debt crisis was avoided. This is because non-oil exports surged in response to a more competitive exchange rate, pushing down the debt/GDP ratio to a manageable 30 per cent by 1990 (Bhattacharya and Pangestu 1993).

The export surge reflected successful economic stabilization together with industrial policy reform. Whereas protection of the manufacturing sector had intensified during 1982–5 in the wake of the second oil boom, thereafter a start was made on dismantling the protective system that had been built up during the oil booms. Non-tariff barriers were substantially reduced, the number of tariff bands was halved and the average level of nominal protection was lowered to 20 per cent between 1986 and 1991. Competition was enhanced by an easing of restrictions on foreign investment, and financial markets were liberalized. Meanwhile, the government cut its share of total investment from 48 per cent to 37 per cent during the 1980s.

The Indonesian reforms of the late 1980s went some way towards returning Indonesia to the East Asian development model. Instead of attempting to speed diversification into HCI, Indonesia resumed its level in the fourth tier of the East Asian escalator, following Malaysia and ahead of Vietnam (Table 10.1). The country's exports had diversified sufficiently for it no longer to be described as an oil-driven economy. The reforms boosted non-oil exports to 74 per cent of total exports in 1991, manufacturing alone accounting for 45 per cent. Within manufacturing, textiles overtook plywood in importance, to account for one-seventh of total exports, while RBI ranked third (Hill 1990).

Indonesian investment, investment efficiency and GDP growth all recovered in the late 1980s: GDP grew at more than 6 per cent during 1986–91, while the ICOR improved to a modest 4.8 compared with an awful 7.8 in 1982–5 (Bhattacharya and Pangestu 1992). The growing importance of a competitive manufacturing sector is underlined by the fact that growth in agriculture, which had acted as a useful cushion after the oil boom, slowed to 3.1 per cent by 1986–91. This reflected the fact that few opportunities remained to extend irrigation and intensify production.

Sustainability of Indonesian growth

But even as Indonesia consolidated its economic success, critics queried its sustainability. Repetto *et al.* (1987) applied environmental accounting to suggest that the per capita GDP growth rate should be reduced by some 2 per cent to allow for the depletion of environmental resources. In addition to the exhaustion of fund resources like hydrocarbons, renewable resources have been depleted as a result of soil erosion and deforestation.

Pearce *et al.* (1990) estimate the soil loss in the densely settled upland watersheds of Java at 10–40 tonnes/hectare per annum. Poor access to credit and markets caused small farmers to deplete the forest cover and to grow crops which were ill-matched to local soil and slope conditions. Fig. 10.4 traces the potentially adverse impacts, including those for lowland padi farmers and city dwellers. Improvement requires a redistribution of farm subsidies away from the lowlands (where they encourage the wasteful

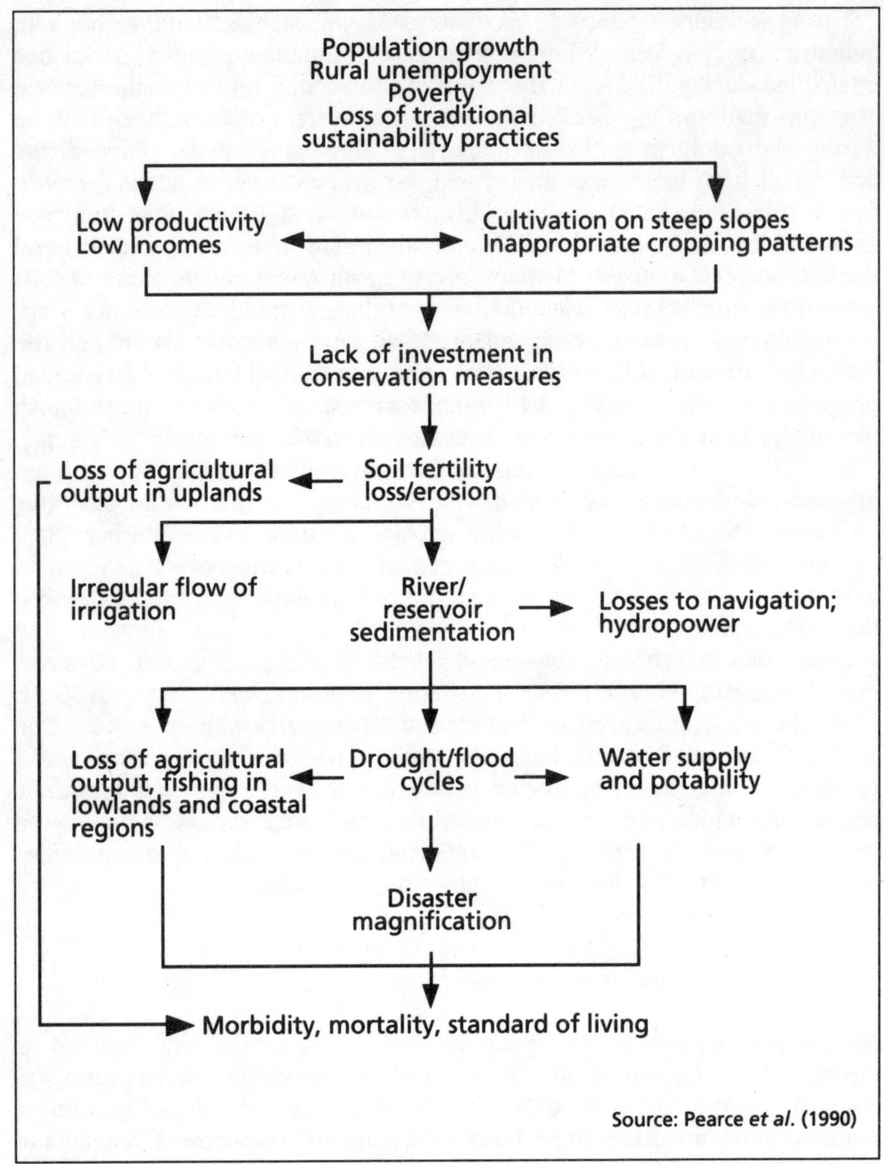

Fig. 10.4 *Indonesia: uplands erosion and its consequences*

application of chemical inputs). This redistribution will provide roads and credit for upland farmers, who will then be able to adjust their land use to more appropriate crops and also to secure more off-farm employment opportunities.

Meanwhile, in the sparsely populated outer islands of Indonesia such as Sumatra and Borneo, where densities are one-tenth or less than 800 per square kilometre of Java, forests are being degraded. As in Brazil, the main cause of deforestation is settlement by small farmers, with some 1.4 million families estimated to have migrated from Java since the 1950s as a result of both government-sponsored transmigration and spontaneous moves. Sustainable development requires that recognition be given to the full value of the forest. This includes its non-timber crops, timber-processing and tourism, as well as wider services such as soil conservation, water run-off regulation, biodiversity promotion and carbon dioxide absorption. More specifically, an economic case can be made for conserving almost half the Indonesian forests. Elsewhere, the government should promote technical assistance for small farmers and provide incentives (such as longer leases and higher stumpage charges) for loggers to practise sustainable forestry.

But as increasing numbers of Indonesian migrate to the cities, so water and air pollution, solid waste disposal and land management become major problems. The annual damage from air and water pollution in Jakarta alone has been estimated at more than $1 billion, with perhaps 10–40 per cent to be added for the costs of time lost due to traffic congestion (Brandon and Ramankutty 1993). Urban transport, which grows with a 7-year doubling rate in Asia, is the main cause of air pollution. Emissions in Jakarta rose five-fold during 1975–88.

The establishment of an environmental protection agency will introduce environmental standards and incentives. Advantage can also be taken of the country's rapid growth, because it means that by 2010, an estimated 85 per cent of its industrial plant will have been built during the preceding twenty years. This gives considerable scope to ensure that the best practice technology is quickly in place. Overall, high-growth Asian countries like Indonesia and Malaysia need to spend between 2 and 3 per cent more of their GDP annually than they presently do on environmental improvement.

Conclusion

Malaysia and Indonesia show how – contrary to the views of Singer and Prebisch and the dependency theorists – primary product export-led growth can lead to self-sustaining economic growth in the post-colonial era. Moreover, both countries have successfully managed ethnic problems, although there can be little doubt that rapid per capita income growth helped, providing governments with the means to cushion potentially disadvantaged groups. Both countries achieved substantial reductions in poverty, with an especially dramatic fall between 1970 and 1990 in the level of absolute poverty in Indonesia, from 60 per cent to 16 per cent (Bhattacharya and Pangestu 1993).

Malaysia and Indonesia also show that a rich natural resource endowment need not inhibit efficient industrial diversification, although the industrial policy errors which both countries made during the oil booms underline the temptation to use the resource bonus to follow lax policies. Both countries tried to force the pace of industrialization during the oil booms (Malaysia through its over-ambitious state-enterprise-dominated HCI big push, and Indonesia through its tolerance of high levels of rent-seeking behaviour), but in neither case was the damage lasting. This is because of the priority given to macroeconomic orthodoxy, which led to the rapid reining in of over-ambitious industrial policies when the oil windfall declined. Although the Indonesian resource endowment has much in common with that of Nigeria, its superior economic performance confirms that the resource curse thesis is not deterministic.

Successful primary product-led growth requires diversification away from a mono-product economy, as export base theory suggests. Competitive industrialization then continues the process of building a flexible and resilient economy. This means that governments must correctly judge a country's position in the development escalator and, if they have adequate administrative skills, they may seek to accelerate emerging competitive advantage. Two countries which attempted to do this without the advantage of a rich natural resource base are Korea and Taiwan. Their example casts interesting new light on the controversy over the role which the state should play in economic development.

11

The role of the state: lessons from resource-poor Korea and Taiwan

As an increasing number of developing country governments seek to learn from East Asian experience, the policy controversy has narrowed down to the question of how much governments need to intervene within a broadly orthodox policy framework (Williamson 1993). This chapter examines the controversy by comparing state intervention in the two largest Asian dragons, Taiwan and Korea. It begins by summarizing the debate between those (mainly orthodox economists) who consider macroeconomic policy to be the critical determinant of successful development and those (the institutionalists) who focus on what governments can do to assist their countries to acquire new industrial technology.

The second section of the chapter sets the international context in which such policies have been pursued. It does so by examining postwar trends in regional industrial comparative advantage. It makes use of an income-driven product cycle model to trace the relatively recent emergence of NIC competitiveness in a widening range of heavy and chemical industry (HCI). The case for strong state intervention to promote HCI received a boost during the 1980s from the new strategic trade theory, and this new perspective is briefly reviewed.

The third section of the chapter compares rates of maturation for two contrasting HCI sub-sectors, steel and automobiles, with reference to Korea and Taiwan and also, where appropriate, to other NICs. It may be recalled that sectoral maturation measures the rate at which infant industries 'grow up' to achieve global levels of technology and competitiveness. The fourth section then compares changes in the structure of the manufacturing sectors in Korea and Taiwan. It relates the structural changes to government policies and explains their impact on economic performance. It is also noted that until very recently, industrial policy neglected environmental issues in both countries. The concluding section summarizes the arguments for strong and weak state intervention.

The industrial policy controversy

In the post-war period, autarkic industrial policies were adopted by market-rich economies (like the large low-income Asian countries and the larger Latin American countries) and also by resource-rich countries such as the oil-exporters. In the case of the larger countries, such policies drove down the share of exports in GDP below 5 per cent compared with more than 30 per cent in the successful mid-sized East Asian countries. Autarkic policies also pushed the effective rate of protection to levels in excess of 100 per cent. The effective protection rate compares the size of the protective tariff with value added only, rather than with the total value of the product (as in the case of nominal protection measures).

Such high levels of protection gave little incentive to manufacturers to emulate international standards. Consequently, sectoral maturation rates often exceeded several decades, whereas a rate of five to eight years is needed if the discounted long-term benefits of the matured industry are to compensate for the initial costs of infant protection (Kreuger and Tuncer 1982). As uncompetitive manufacturing grew in relation to GDP, the economy experienced difficulty in generating sufficient foreign exchange, so that economic growth became both slower and more erratic. Yet reform was frequently blocked, because the system of protection conferred sizeable rents (i.e. returns in excess of those required by an efficient producer to remain viable) on those in the protected sectors. For example, Bergsman (1970) calculated that in Brazil in the late 1960s, rents totalled around 9 per cent of GDP. Unable to implement reform, many governments turned to 'growth-based' solutions which attempted to use public spending to renew economic growth. These 'populist booms' sometimes took the form of HCI big pushes. Such policies led in the 1970s to the imprudent accumulation of foreign debt, which subsequently proved difficult to service in the 1980s, as discussed in Chapter 10.

The IMF and World Bank attempted to use their lending policies during the 1980s to unwind the cumulative distortions in the economy which the pursuit of autarkic industrial policies had created. The loans for 'structural adjustment' carried conditions which aimed to reduce state intervention. The IMF called for cuts in public expenditure and also for exchange rate devaluations, in order to redeploy capital and labour into more competitive activity. The reforms led to a radical shift of resources away from the erstwhile protected sectors and so imposed sizeable hardships (Selowsky and van der Tak 1986). Not surprisingly, they proved politically difficult to implement, a problem which was not helped by the sometimes insensitive attitude of the lenders, notably the IMF, towards domestic political constraints.

In backing such harsh reforms, iconoclastic critics of state intervention like Balassa (1982) and Lal (1983) adopted a dogmatic attitude which even attacked the policies of successful industrializers like Korea. Less stridently,

Ranis and Mahmood (1992) attribute East Asian success to the skilful pursuit of prudent macro policies which steadily liberalized the economy. They argue that orthodox macro policies reduce the scope for the 'covert' transfer of resources between groups in society. Such covert transfers include the rents which both owners and workers in the protected manufacturing sector and associated service sector enjoy at the expense of the politically weak rural poor in particular, but also at the expense of domestic consumers of manufactured goods in general.

Yet institutionalists like Amsden (1989) and Wade (1990) defend government intervention in East Asia, and their stance is backed by the Japanese directorate at the IMF. The Japanese feel that, whatever the capacity of other developing regions, East Asia clearly shows that state intervention can successfully promote rapid structural change. For example, as their economies reached the turning point, countries like Korea and Taiwan which pursued competitive industrial policies (Table 5.1) conferred time-constrained incentives on sectors of emerging competitive advantage such as heavy and chemical industry. Most importantly, they did so without weakening the competitiveness of existing competitive sectors, mainly light industry.

Yet the incentives under a competitive industrial policy differ little from those of an autarkic policy. They both include tariff protection, tax holidays, low-interest loans and help with technology acquisition and marketing. The key difference is that, under a competitive industrial policy, the incentives are given in return for the *rapid* achievement of economic and technical maturation, and they are quickly phased out. In the case of Korea, the recipients of incentives were often penalized if agreed targets were not met (Song 1990). In this way, the scope for rent-seeking behaviour is constrained and the growth of manufacturing does not become a burden upon the rest of the economy.

Wade (1990) argues that the time-constrained incentives of a competitive industrial policy reduce the risk to investors by, in effect, socializing the risk of entry into new industries. Amsden (1989) agrees, and forcefully argues that, contrary to orthodoxy, Korean success owed much less to 'getting prices right' (the orthodox slogan) than to deliberately 'getting prices wrong'. She considers that this was necessary in order to stimulate investment in new manufacturing, which would otherwise not occur because of the perceived risks in making large investments with lengthy pay-back periods. But although Amsden and Wade make a plausible case for the pursuit of an active industrial policy, neither provides a definitive proof.

Meanwhile, the orthodox economist Stern (1990) has statistically analysed structural change in Korea and found no evidence that it occurred faster than would have been the case without an active industrial policy. Ranis and Mahmood (1992) concur, and consider the Korean HCI big push to have been a policy relapse in what was otherwise a 'linear' liberalization (i.e. constant steady opening) of its economy after 1963, in line with orthodox prescriptions. They compare the aggressively interventionist policy of Korea unfavourably with the more cautious Taiwanese policy. Korean policy was certainly strongly criticized in the mid-1980s (Park 1986, World Bank 1987).

Yet by the late-1980s, a strong export-led rebound by Korean HCI combined with a sharp slowdown in Taiwanese investment to cast doubt on the greater merits of the Taiwanese approach. Moreover, the emergence of strategic trade theory championed by Krugman (1986) during the 1980s suggested that many global markets are characterized by imperfect competition. This view echoes that of dependency theorists (see Chapter 8). Imperfect competition provides the rationale for able governments to intervene to enhance national welfare. But before comparing government intervention in Korea and Taiwan, it is important to establish when, why and how the shift in manufacturing competitiveness towards the NICs began. An income-driven product cycle model, based on the pioneering work of Vernon (1966), helps to identify and explain the change.

Product cycle, trade theory and comparative advantage

The product cycle

Vernon's (1966) product cycle model is driven by technological innovation, which initially encourages location close to major markets. Location within such markets allows producers to capture the external economies and thereby reduce their risks as new entrants. But as the production technology becomes standardized, so manufacture relocates to regions with lower labour costs. Vernon (1979) felt the model explained postwar trends within the OECD countries as the lagging countries (Japan and Western Europe) attracted more sophisticated industries as they steadily closed the technological gap on the United States.

Stobaugh (1970) adapted the product cycle model to HCI, which is often characterized by economies of scale. He concluded that changing market size, rather than technological evolution, drove locational change. He explained the initial location of HCI at the major market in terms of the adequacy of demand in such markets to support a plant of minimum viable size. Lagging regions with smaller sub-optimal markets were more effectively served by imports. But, as demand in the lagging markets expanded and crossed the minimum threshold for viable plant size, production diffused to them (Table 11.1). This diffusion of scale-sensitive plants to emerging markets was driven by the need to minimize freight costs.

Stobaugh's scale-based reformulation of the product cycle model can be elaborated into an income-driven product cycle model. This is done by formalizing the link between three variables: per capita income in the emerging market; the stage of the market within the product cycle; and competitiveness.

Table 11.1 *Revised product cycle model*

Stage	Pioneer	Dynamic	Maturity	Eclipse
1 Income elasticity	High	High/falling	Falling	Low
2 Growth of demand	Erratic	High	Slowing	Slow/negative
3 Dominant production factors	R & D	Market, cheap labour	Skill, capital	Subsidy
4 Market structure	Oligopoly	Declining oligopoly	Perfect competition	Concentration/state monopoly
5 Investment risk	High	Declining/low	Rising	High
6 Corporate product strategy	Single product	Single product	Dominant product	Diversification/subsidized single product
7 Corporate trade stance	Domestic only	Domestic + export	Domestic + foreign subsidiary	Protection/import

Source: After Tichy (1989, p. 5).

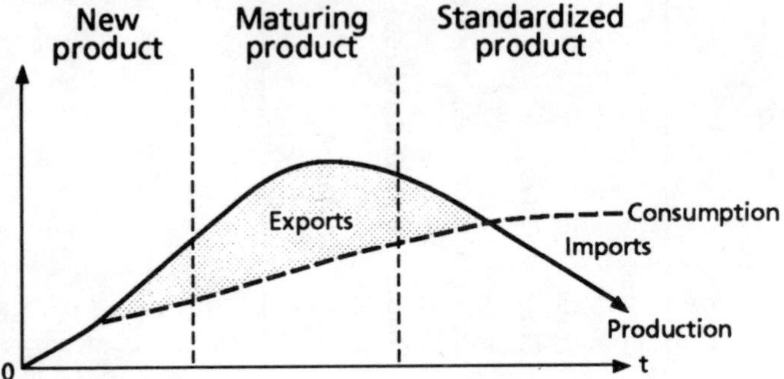

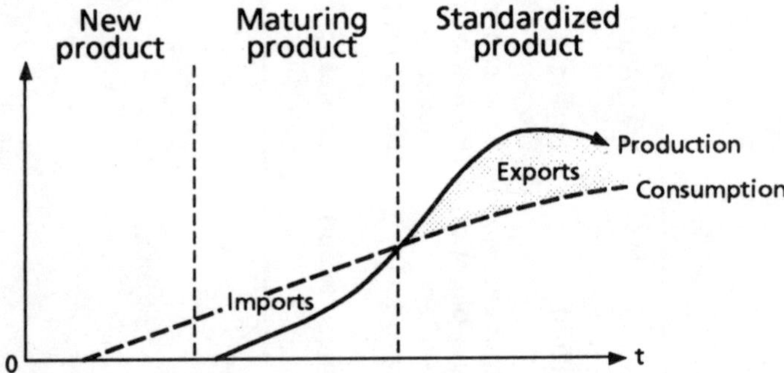

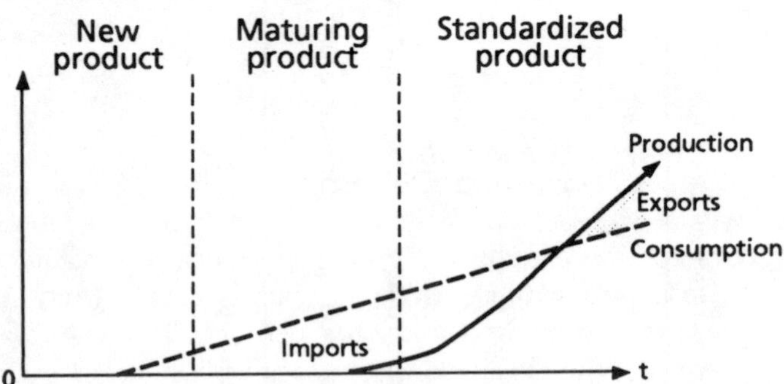

Source: Williamson and Milner (1991)

Fig. 11.1 *Product cycle location trends*

The income-driven product cycle model assumes that the demand curve for a product is S-shaped (Fig. 11.1). A pioneer stage of low demand gives way to a high-growth youthful stage (in which demand for the product outstrips growth in GDP) as rapid market penetration occurs. Thereafter, market saturation causes growth to decelerate to a rate at, or below, that of GDP, and demand passes into the mature stage of the product cycle. The appearance of substitute products may even add an eclipse stage. Table 11.1 traces the macroeconomic implications of the model in lines 1 to 4 and the micro, or corporate, aspects in lines 5 to 7.

Some evidence for this cycle is provided for steel. With rising per capita income, the amount of steel used per $1000 of GDP per capita at first rises steeply and then declines as demand shifts from goods to services and a wider range of materials is used (Bernstam 1991). In the case of automobiles, Karmokolias (1990) notes that below a per capita income level of $2000 (in 1980 dollars), demand is slow-growing and there are typically only two vehicles per 100 people. As incomes rise above $2000, to $4000, however, demand accelerates, so that ownership nears forty vehicles per one hundred people. Above $4000, incomes approach those of the industrial countries and demand is saturated.

Applying the income-driven product cycle model at the global level, differing regional markets are at differing stages of the cycle, depending on their level of economic development (Fig. 11.1). All else being constant, the income-driven product cycle model states that producers located in large markets in the youthful stage of the product cycle have a competitive advantage over those in markets in the pioneer, mature or eclipse stages. This is because the mechanism that drives changing locational advantage in this model is not technological change (as in Vernon's original model), nor market size alone (as in Stobaugh), but a growth dynamic. Briefly, the growth dynamic (Verdoorn effect) enhances producer competitiveness in fast-growing markets, whereas its absence in mature or eclipse stage markets weakens the competitiveness of producers located there.

One explanation of the Verdoorn effect is the impact of 'learning by doing': the greatest productivity gains are to be made in the early stages of the product cycle; productivity gains subsequently taper off as markets mature. However, while this may reinforce the growth dynamic in the more successful NICs, another factor at work is a virtuous circle of growth effects (Auty 1987, Schorsch 1989). Fast-growing markets lower the risk of investment in large-scale projects, by reducing the likelihood that capacity will remain surplus to needs for any length of time. Fast growth also implies that the average factory is relatively new, so that productivity is high. This, in turn, makes it easier for producers to concede wage demands to attract trained workers or to gain worker cooperation, compared with producers in slow-growth mature markets.

In contrast to such advantageous conditions in dynamic youthful markets, those markets in the mature stage of the product cycle experience decelerating demand (Table 11.1). Such markets are supplied at least risk either by imports or by capacity creep (the squeezing of more output from existing plants). Producers located in such markets tend to skimp on capital investment, so that the average age of plant increases and productivity may lag. There is a risk that such factories will fossilize and be run into the ground (Auty 1975).

Meanwhile, workers in mature markets often prove reluctant to adjust their work practices, further eroding competitiveness. This occurred, for example, in the US integrated steel industry in the 1960s and 1970s (Crandall 1981). Applying the product cycle model at the global scale, the competitive advantage in HCI switches from those producers in markets where demand is decelerating, to those in markets with lower per capita income which have crossed the minimum viable plant threshold size and are in the dynamic (youthful) phase of the product cycle (Fig. 11.1). This process implies a loss of competitiveness by the industrial countries to the NICs, and eventually by the latter to the next tier of industrializing countries.

During the postwar period, however, increases in the size of the minimum viable plant delayed the shift in comparative advantage in steel production and automobile assembly to the emerging NICs. The markets of even the largest NICs lacked sufficient demand to support a steel plant of minimum viable size until the 1970s and to support a car assembly plant until the 1980s (Womack *et al.* 1990, UN 1950–85). More specifically, the minimum viable size for an integrated steel mill rose from around 500 000 tonnes in the mid-1940s to 1 million tonnes by the mid-1950s, and 5 million tonnes by the early-1980s (Barnett and Schorsch 1983). As for automobile assembly, the largest world producers each made in excess of 2 million units by the 1980s. Although the minimum viable size of an assembly plant remained around 200 000 vehicles, that for engines and transmissions was twice as big. As a result, a firm producing a market-spanning range of cars in plants of minimum viable size required an annual volume of at least 1 million units (Lucke 1988).

In the context of the developmental escalator, discussed in Chapter 10, a successful industrial policy needs to correctly identify emerging comparative advantage. It needs to encourage efficient investment in such sectors without damaging sectors of existing (but waning) comparative advantage. The East Asian NICs were far more successful in doing this than the larger Latin American and Asian countries, whose entry into HCI was invariably premature. Meanwhile, the emergence of the new strategic trade theory during the 1980s appeared to confirm the wisdom of state intervention to assist emerging sectors, provided the timing was correctly judged. This appeared to strengthen the interventionist case.

Strategic trade theory

Prior to the 1980s, trade theory was based on the notion that the pattern of trade between countries is determined by differences in their relative endowments of factors of production (i.e. of capital and labour). It offered powerful reasons why trade occurs between rich and poor countries. Yet as an increasing number of industrial countries converged on the US capital- and skill-intensive factor endowment during the postwar period, trade between them increased, rather than decreased. Expressed another way, intra-industry trade (trade involving similar products) was becoming more important than inter-industry trade (trade involving the exchange of dissimilar products). This called into question the previous assumptions of trade theory, including one of its hitherto most robust conclusions, namely that free trade between countries is invariably mutually beneficial.

The new strategic trade theory suggests that global market imperfections (due largely to the presence of scale economies) create rents whose allocation among countries may be influenced by government intervention which creates national champions that capture a fraction of such production (Brander and Spencer 1985). The new trade theory also suggests that external economies affect trade and that government intervention to promote industries subject to external economies, such as automobile assembly, will raise living standards in their countries at the expense of other countries whose governments do not so intervene (Krugman 1993).

The apparent legitimization of government intervention in support of key industries by the new strategic trade theory was especially timely for some industrial country governments (the US and those of the EC) during the 1980s because of the problems posed for their economies by the slowdown in the global economy. These governments intervened to 'restructure' ailing industries. As a result, whereas successive global trade agreements had reduced the average level of tariffs on manufactured goods from 40 per cent in 1947 to less than 5 per cent by 1990, the share of industrial country imports subject to non-tariff barriers (the 'new protectionism') rose slightly, from 15 per cent to 18 per cent between 1981 and 1986 (Crook 1990).

But doubts have grown about the new trade theory. First, the direct benefits of government intervention are difficult to assess, because they entail heroic assumptions about the size of the externalities (Krugman 1992, Stewart and Ghani 1992), the speed of corporate maturation and the level of rents to be secured on domestic and export sales. Meanwhile, as earlier chapters have shown, the direct costs of support fall principally on domestic consumers in the form of restricted product choice and higher prices. Finally, the resulting transfers of rents from domestic consumers are not guaranteed to be ploughed into investment, because their deployment is often contested by trade unions. There is evidence from the United States and the EC that workers in protected markets secure returns in excess of those achieved in more competitive sub-sectors (Dixit 1988). Such groups then form an important political constituency, along with their weakened employers, lobbying against trade and market liberalization.

More recent research on strategic trade theory also casts doubt on the value of such intervention. By the 1990s, three strong conclusions had emerged which tend to reinforce the wisdom of a commitment to free trade.

First, although net benefits may be won through strategic trade policy, they are likely to be small and to be measured in fractions of 1 per cent of GDP (Dixit 1988). Second, the probability of errors by intervening governments and of reprisals by aggrieved trading partners creates a sufficient threat to the stability of the global trading system to render the very modest potential gains from strategic trade intervention not worth the risk (Krugman 1993). Third, strategic trade intervention may prove counter-productive due to misjudgement, as US efforts to restrict Japanese car imports in the 1980s show (Bergsten and Noland 1993).

In the context of the emerging (albeit delayed) NIC competitive advantage predicted by the product cycle model, and also of the waxing and waning of the theoretical case for state intervention in trade and industry policy, NIC sectoral maturation rates are now explained for one capital-intensive HCI sector (steel) and one skill-intensive HCI sector (automobile assembly).

Sectoral maturation rates

Integrated steel production

The governments of both Korea and Taiwan intervened to build up domestic manufacturing, but the Korean intervention was the most pervasive. The World Bank rejected Korean plans to build an integrated steel plant in the 1960s, on the grounds that the country lacked a large home market, domestic raw materials and skilled labour. But Korea persisted with its plan and used war reparations from Japan to construct a world-scale steel plant. It established a state-owned steel company, Posco, but gave it more commercial autonomy than most public enterprises. One important consequence of state ownership was that the pressure which private shareholders exert for dividends was diminished. This favoured the rapid expansion of investment, as did cheap capital, tax holidays, accelerated depreciation, assistance with infrastructure construction and tariff protection.

Posco executed eight expansions in two plants between 1973 and 1992, which raised its rated crude steel capacity to 20 million tonnes – around three-quarters of total domestic capacity by the late-1980s (Table 11.2). Posco set high initial standards for the equipment it purchased and the effectiveness of its operation. Posco modelled itself on Nippon Steel, from which company it received initial technical assistance which was, however, quickly phased out (Amsden 1989, Enos and Park 1988). Posco set tight targets for the construction and start-up of each stage, and steadily reduced the completion times as its expansion proceeded (Table 11.2). The rapid rate of project completion, when combined with the financial incentives and a cheap skilled labour force, cut the costs of construction well below those of competitors. Dornbusch and Park (1987) argue that Korea's adaptable labour force is the cornerstone of the country's growth.

Table 11.2 *Capital outlays for steel plant construction*

Plant	Phase	Capacity	Cost/ tonne	Date completed	Months building
Posco	1	1.03	287	6/73	39
Pohang	2	1.57	352	5/76	29
	3	2.90	469	12/78	28
	4	3.60	386	2/81	24
Posco	1	2.70	723	5/87	26
Kwangyang	2	2.70	473	7/88	22
	3	2.70	1163	12/90	23
	4	3.30	863	10/92	21
CSC	1	1.50	1072	12/77	39
Kaohsiung	2	1.75	773	10/82	53
	3	2.40	510	4/88	45

Memo item: Total plant capacity, total investment, average contruction cost/tonne

Posco Pohang	9.1 million tones for	$3.59 billion at	$395/tonne
Posco Kwangyaung	11.4	8.89	780
CSC Kaohsiung	5.7	4.22	740

Sources: Paine Webber (1991) (unpublished data), CSC (1992) (unpublished data)

The average construction cost for the four-stage Pohang steel plant was $395 per tonne capacity, scarcely one-quarter that of state-assisted plants in Brazil and Pakistan and only two-thirds as much as a contemporary new Japanese plant, or barely one-third that of a United States plant (Posco 1988). Posco's production costs quickly fell and the plant became internationally competitive by 1981, within 8 years of start-up. Such a maturation rate is just within the range needed to justify infant industry support, according to the neo-liberal school of thought. Meanwhile, in Taiwan, state-owned China Steel followed more cautiously – later, and on a smaller scale and at a slower pace than Posco. Taiwan achieved similar technical results, but at less financial cost. Elsewhere, under autarkic policies in Brazil, Mexico, India and China, maturation rates were orders of magnitude longer.

The rapid accumulation of steel assets and the expansion of sales by Posco was associated, however, with a relatively disappointing overall rate of financial return. The outcome is the more disappointing because of the substantial scale of state assistance given (which Japanese producers estimate was equivalent to 20 per cent of production costs). The very low real return on capital implies high opportunity costs and requires large long-term benefits, to be justified. Yet such negative assessments must be qualified by the fact that the global steel industry underwent an unexpected decline in profitability from the mid-1970s, after the commitment to a Korean steel industry had been made.

Car assembly

The postwar efforts of the NICs to establish auto production were invariably premature. This is because, prior to the 1980s, no developing country had sufficient domestic sales to support one producer of minimum viable size. Earlier entrants would therefore need to export in order to capture the economies of scale. Korea and Taiwan mistakenly attempted to establish car assembly from the early 1960s, 2 decades too soon. The Korean car market only entered the dynamic phase of growth (when domestic demand crosses the minimum size threshold and the product cycle model predicts that competitive advantage will strengthen) in the mid-1980s, while the smaller Taiwanese market was still further behind.

The Taiwanese car sector did not recover from the protectionism which its premature start demanded. It remained a classic case of industrial policy capture by vested interests (like the car sectors of Brazil and Mexico), until reforms began in the late-1980s. The Korean outcome is less clear-cut: if the attempt to establish a competitive car sector is traced from the mid-1970s, it can be argued that maturation was achieved within a decade.

The new Korean car industry policy of 1974–6 was part of the HCI big push. Korean success lies in tight control of the number of vehicle producers in order to limit the dissipation of the economies of scale among too many new entrants. Three car assemblers were established, each linked to an MNC: Daewoo (GMC), Hyundai (Mitsubishi) and Kia (Isuzu). High and persistent tariff protection was maintained, along with restrictions on component imports. The Korean incentives were more generous than those in Taiwan, but the penalties for non-performance were also harsher. The latter appears to be the principal reason why Korea had much more success than Taiwan in maturing auto assembly. Expressed bluntly, the Korean government was a more effective bully than its Taiwanese counterpart (Song 1990).

Despite the control on new entrants, Korean domestic demand was still inadequate. The largest company, Hyundai, responded by reversing the normal marketing sequence (from domestic-driven demand to export-driven demand) and built an export-oriented plant. The circumstances of this risky strategy were fortuitous, however, partly because of a gap in the North American market that was created by a Japanese agreement to restrict its share of that market. The Korean car export drive of the mid-1980s was threatened, however, by a technological lag in the components sector, which adversely affected quality.

Table 11.3 traces the maturation of the Korean automobile industry. Hyundai, the leading Korean assembler, was set up in 1968 as a joint venture with Ford, assembling kits with 30 per cent local content. It launched the first Korean car in 1976, after failing to reach agreement with Ford. The new car was designed 50 per cent in-house, with the gaps filled by Italian body designs, Japanese transmission and British parts (Stern 1990). In addition, British management skills were secured for the model start-up during 1974–7

Table 11.3 *Development stages in the Korean automobile sector since 1962*

Stage	Foundation 1962–74	HCI push 1975–81	Export drive 1982–8	Home base 1989
Character	Kit assembly	Local model, mass production	Restyling, JIT, front-wheel drive	Advanced design of engines and transmission
Local content	30%	85%	97%	97%
Models produced	9	11	10	13
Mid-stage volume	14 000	57 000	264 000	>1 000 000
Technology source	Licensing, joint venture	Licensing, joint venture	Licensing, joint venture, in-house R&D for exports	Increasing in-house R&D
Acquired technology	Inspection, production management	New model development, quality control	Mass production JIT, front-wheel drive, US standards	Design with advanced technology

Source: After Hyun (1989)

through the appointment of six technical experts and a Leyland executive as vice-president (Hyun 1989).

In 1978 Hyundai established its own research and development centre at Ulsan with the assistance of Mitsubishi. It launched its first wholly in-house designed car (the Excel) in 1985 and acquired an impressive technological capability compared with other NIC producers. Korea's rate of model replacement matched OECD levels and its level of automation exceeded that of other NICs. Nevertheless, as late as 1991 Hyundai complained that its products experienced three times the faults of Japanese cars.

But the costs of Korean entry into auto assembly were high: the effective rate of protection on transport equipment was more than 300 per cent by 1978, compared with a manufacturing sector average of 30 per cent (Young and Yoo 1982). Korean exports required large subsidies until the mid-1980s. One Japanese source estimated that the subsidy on Korean car exports was equivalent to $1000 per vehicle (Cho 1989). The principal subsidy was the right to assemble one imported car kit (usually a large luxury car, commanding a high price because of repressed demand) for every five vehicles exported. The costs of this subsidy were therefore borne by the Korean consumer in terms of higher vehicle prices.

Summarizing, Korea set best practice in entering new industries with sectoral maturation rates close to those required by orthodox theorists. Maturation took less than a decade in the case of both steel and auto assembly, less than half the time taken by Brazil and Mexico whose industries also regressed. There is clear evidence that both Latin American countries could have matched Korean economic maturation rates, albeit with greater reliance on external technical assistance in the case of automobiles. To achieve this they would need to have adopted both orthodox macro policies and a competitive industrial policy, as the dramatic improvement in the Mexican car industry in the late 1980s shows.

However, the costs to Korea have been high, not only in terms of subsidies, but also in terms of the negative effects which the 1974–9 HCI big push had upon macroeconomic performance. The less interventionist strategy of Taiwan (which recognized its own failings due to premature entry into auto assembly and, unlike Korea, sought to cut the losses), therefore has much to commend it.

The macroeconomic impact of industrial policy

Shared Korean and Taiwanese preconditions

Taiwan and Korea shared important socio-economic characteristics in the immediate postwar years. They both experienced civil war, in the late-1940s (China) and early-1950s (Korea). Each country had a large influx of refugees which created initial problems of labour absorption. In the case of Taiwan, an added difficulty was the potential conflict between the 2 million mainland Chinese refugees and the 6 million local inhabitants.

In addition to the shared experience of coping with refugees (an extra source of cheap labour, as well as more mouths to feed), both countries secured three advantages for industrialization as a result of having been Japanese colonies in the first half of the twentieth century. One advantage was the accumulation of industrial infrastructure, and the second was experience of manufacturing, although the economy of each country remained predominantly rural (the partition of Korea left most of the heavy industry in the communist north). The third advantage from the Japanese occupation was progress in land reform, particularly in the case of Taiwan.

Taiwanese progress in land reform was consolidated in the immediate aftermath of the Japanese withdrawal, whereas land reform in Korea at first lagged. Early in their occupation of Taiwan, the Japanese had removed the large landlords and encouraged farm production changes. This had led to a rise in per capita income of around 1.5 per cent annually between 1895 and 1940, and some real improvement in the standard of living for the average Taiwanese (Ho 1978). The additional land reform during 1949–53 established a large number of small owner-occupiers, whose output grew by 8 per cent annually over the next decade (Amsden 1985).

During the 1950s, the dynamic Taiwanese agricultural sector created a fast-growing domestic market and also yielded substantial transfers to support its primary industrial import substitution phase. These transfers persisted into the late-1960s and were achieved by setting high prices for state-controlled fertilizer supplies, and by compulsory purchases of about half the rice crop (Ho 1978). Meanwhile, the displaced Taiwanese landlords had been encouraged to set up urban businesses with the financial compensation they received for their land.

In contrast to Taiwan, the productivity of Korean agriculture stagnated until the 1960s (Ho 1981). Song (1990) argues that the benefits of the rapid economic growth rate under the Japanese occupation (almost 4 per cent per annum) accrued largely to the colonizers. For example, half of Korea's rice production was exported in the 1930s, and Korean per capita income actually fell during the inter-war years, whereas that of Taiwan increased. During the 1950s, the upheaval of land reform in Korea combined with unsound macroeconomic policy to mute rural growth. The Korean government was preoccupied with import substitution industry, which was inefficiently executed, and with universal secondary education, which subsequently paid handsome dividends.

Land reform did improve Korean income equality, however, albeit more belatedly than in the case of Taiwan. This was an important precondition for the country's post-1963 high-growth phase (Suh 1992). But during the 1950s, Korean agriculture could not match that of Taiwan in providing a flow of foreign exchange, labour, and capital (Ka and Selden 1986). Overall, the

Korean economy during the 1950s was less resilient than that of Taiwan, and more dependent upon external infusions of capital. Korean per capita GNP grew by only 0.7 per cent 1954–62, well below the Taiwanese rate of over 5 per cent.

Policy differences

Taiwan had a narrower range of industrialization options than Korea, because of its smaller domestic market and limited geographical area (using area as a proxy for natural resources range). In the early-1950s, Taiwan had a population (8 million) one-half the size of Korea in an area (36 000 square kilometres) one-third as large. Certainly, Liang and Liang (1988) argue that Taiwan's small size is an important reason why the government abandoned its autarkic industrial phase earlier than did Korea.

Unlike the small but resource-rich countries of Latin America or sub-Saharan Africa, Taiwan had neither the mineral resources nor the agricultural potential with which to continue to support an overly protected industrial sector. Moreover, the shortage of cultivable land, together with external controls on the principal exports of sugar and rice, meant that agriculture held little long-term prospect of sustaining the rapid growth in the economy. Yet Taiwan's domestic market for manufactured goods was too small to allow for rapid industrial diversification. Consequently, export-oriented manufacturing (perhaps modelled on Hong Kong) was left as the most practical option (Tsiang 1984).

Whatever the reason, Taiwan liberalized its trade regime earlier than Korea, opening up its economy some 5 years before Korea (Table 11.4). The reduction of import barriers and a large devaluation in 1958 heralded a spectacular phase of export-led growth in Taiwan. The trade liberalization of the late 1950s triggered a rapid expansion of manufactured exports, whose share of total exports jumped from 10 per cent in 1955 to 46 per cent by 1965.

Taiwanese macroeconomic policy, like its industrial policy, was more cautious and less interventionist than that of Korea (Wu 1988). Taiwan adopted its cautious macro policy in 1958 by adding fiscal balance and (unusual at the time) a strongly devalued currency to the unfashionably high real interest rates which it had adopted earlier in that decade (Tsiang 1984). Korea adopted a more liberal trade policy in 1963, following a strong devaluation, and it too experienced rapid export growth (Wu 1988). Table 11.4 traces the policy shifts which, in line with the East Asian development model and the concept of the development escalator (Table 10.1), moved from labour-intensive exports in the 1960s, to HCI in the 1970s (with a trend from capital-intensive industrial intermediates to skill-intensive engineering) and then to knowledge-based products in the 1980s.

Structural change proceeded apace, with HCI outstripping light industry's contribution to GDP by 1973 in Taiwan and by 1977 in Korea. Rapid economic

Table 11.4 *Industrial policy phases, Taiwan and Korea*

Phase			Timing		Comments
			Taiwan	Korea	
Autarkic			1950–8	1952–62	
Competitive:	1	Export-Led	1959–71	1963–73	Turning point c.1969 Taiwan, 1973 Korea
	2	HCI-Led	1972–81	1974–82	Korea switched to big push 1977–9
	3	Liberalization	1982–91	1982–91	Post-86 exchange rate appreciation

growth resulted in both countries from high rates of productivity growth combined with high levels of investment (see Table 1.5). Following their outward-opening, the per capita income in both countries grew with a less than 8-year doubling rate.

Closer inspection reveals, however, that the more interventionist policy of Korea was associated with a macroeconomic performance which was more erratic than that of Taiwan and which made less efficient use of investment. Inflation in Korea averaged 12.7 per cent between 1961 and 1990, compared with only 5.6 per cent in Taiwan. As for investment efficiency, the Korean ICOR was higher (i.e. less efficient) than that of Taiwan, albeit only slightly so through the export-led phase (Table 1.5). Thereafter it deteriorated to 4.1 during the HCI-led phase (one-third higher than Taiwan) when, in contrast to Taiwan, Korea ambitiously accelerated its HCI drive into a big push.

Divergence: the Korean big push

The Korean HCI big push was associated with accelerating inflation, burgeoning debt, a marked deceleration in economic growth and a large deterioration in the trade balance. More specifically, subsidized credit was used to create excess HCI capacity, which at first gave a low financial return. At the macro level, the HCI big push triggered inflation, worsened the current account and accelerated debt accumulation (Park 1986). The 1979–81 economic recession is widely blamed on the HCI big push, and it triggered radical reforms of industrial incentives, trade and finance. The ending of the HCI big push has been credited with the improved performance of the Korean economy during the 1980s (World Bank 1987).

The Korean decision to proceed via a big push meant that project sequencing was heavily compressed in the late-1970s. The tight bunching of the HCI investments reflected an attempt by the planners to capture the internal and external economies of scale through the synchronization of many large complementary projects. But domestic project implementation capacity was strained during a time of increasing macroeconomic instability. The investment in HCI proceeded much faster than planned. The Korean big push construction boom of 1974–9 raised the share of investment in GDP from 18 per cent to 30 per cent, and increased economic growth to 10 per cent. But the efficiency of investment fell sharply, the ICOR deteriorating from 2.5 to 5.5 through the late-1970s.

As with big pushes elsewhere (see Chapters 5 and 6), the Korean big push triggered rapid inflation. The real exchange rate appreciated and exports lagged as capital-starved traditional export industry (e.g. textiles) lost competitiveness. The country was forced to deflate the economy at the same time as the second oil shock hit: output contracted sharply between 1979 and 1980. Korea borrowed heavily to cover the widening balance of payments gap caused by higher oil prices, HCI machinery imports and weakening export performance.

The post-1979 reforms led the Korean government to cut its role in economic targeting and to concentrate on improving the efficiency of capital markets. The industrial conglomerates (the *chaebol*, like Samsung, Hyundai and Daewoo) which had been the prime recipients of government assistance were by then resented, being accused of reversing the country's traditionally equitable distribution of wealth. Efforts were therefore made to encourage Taiwan-style small and mid-sized firms. Competition was also increased by lowering import tariffs and opening the domestic market up to imports.

A strong economic rebound in the mid-1980s pushed Korean GDP growth to 10 per cent/year during 1986–8, while debt fell absolutely. The composition of exports changed away from textiles and towards electronics, steel, ships and vehicles. The economic success was attributed by the critics of the big push to policy reform, but it also reflects an improvement in external markets (Corbo and Nam 1988) and the utilization of spare HCI capacity. The HCI rebound created a large cash flow for additional low-cost HCI capacity expansion. Nevertheless, it remains far from clear that the Korean big push justified the resources expended, or that the economy would have performed less well without it.

Environmental issues

The industrial policy also provided an opportunity to restructure the space economy of both Taiwan and Korea by establishing HCI at growth poles away from the capital city (Auty 1990b). The Korean growth poles were located mainly in the south-east of the country, within 130 km of the second city, Pusan. The principal industrial poles were at Pohang and Kwangyang for steel, Ulsan and Yochun for petrochemicals, and Ulsan and Changwon for

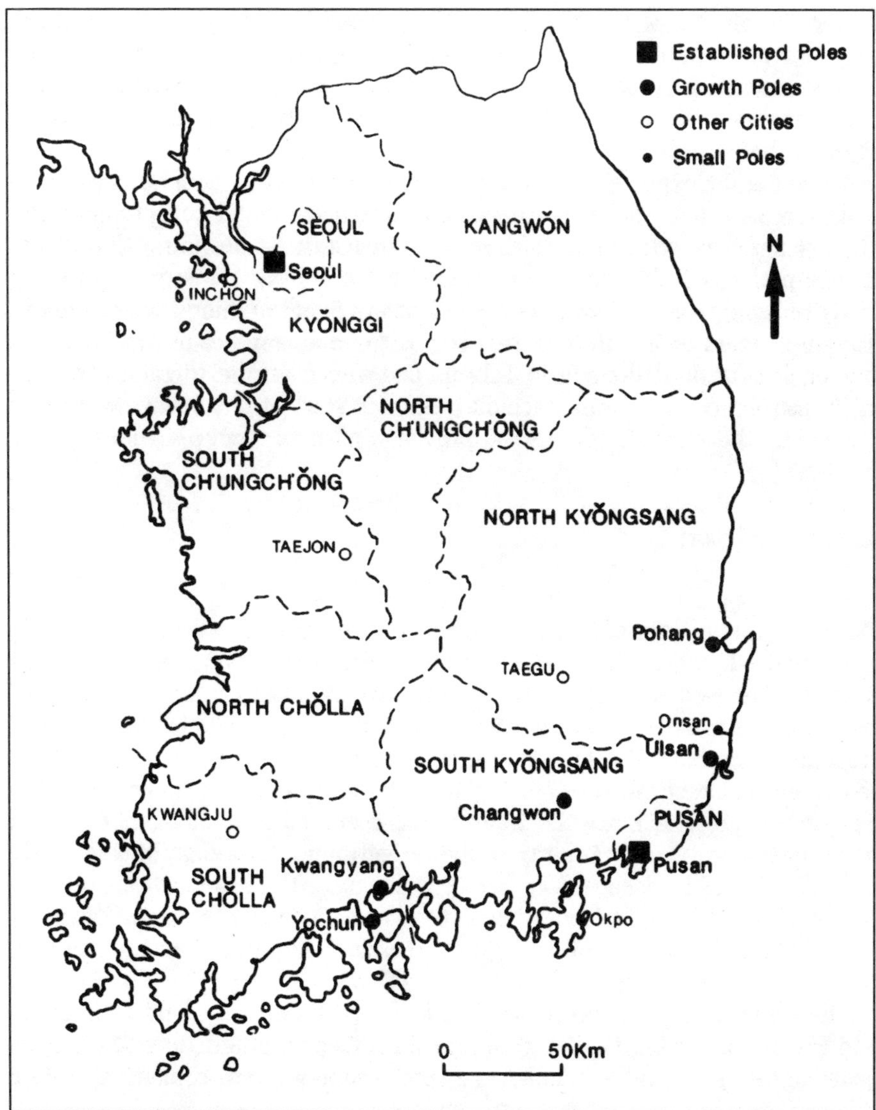

Fig. 11.2 *HCI growth poles in South Korea*

engineering (Fig. 11.2). Similarly, in Taiwan, the southern city of Kaohsiung was developed as a heavy industry growth pole, to counter the dominance of the capital city, Taipei.

The Korean experience suggests that, to some extent, the growth poles redistributed economic activity at the expense of the large but geographically constrained southern city of Pusan. Meanwhile, by dispersing the potentially high-polluting HCI plants, the build-up of industrial pollutants to levels where they provoked public protest was delayed. In neither country did environmental concerns rise strongly to the surface until the 1980s; when they did, their expression coincided with the freeing up of the political environment. Prior to that, a prominent Korean scholar had been hounded by the military government for expressing concern about the quality of water supplies.

As a result of the delayed emergence of environmental concerns, neither country has developed a comprehensive framework for dealing with environmental problems. Although the Korean stress on heavy industry might seem likely to have made that country's problem worse, if anything the situation in Taiwan appears to be inferior. This seems to have resulted in part from the higher density of settlement in Taiwan, and also from the frugal investment in infrastructure in Taiwan (including the late start on a subway system for Taipei and the woefully inadequate provision of urban water supplies).

Conclusions

Whatever their differences, the two resource-poor, market-modest NICs industrialized more successfully than their larger and better-endowed counterparts. Both Korea and Taiwan abandoned autarky relatively early and built on their comparative advantage in labour-intensive exports, before targeting HCI after the labour market reached its turning point. A stable macro environment, with its commitment to fiscal balance, and a competitive exchange rate, has been the bedrock of their success. In addition, both governments gave early priority to the redistribution of assets (mainly land) and also education (especially primary education). These two aspects, in combination with rapid labour-intensive growth, contributed significantly to the rapid elimination of poverty and to a relatively equitable income distribution.

Their success has not been based on *laissez-faire* policies, as Balassa (1982) and Ranis and Mahmood (1992) imply, but rather on pragmatic and effective government intervention. Following the abandonment of autarky, they stressed the development of competitive exports and intervened to encourage this. After the turning point had been reached, they began targeting specific sectors, a policy which had mixed results, at best. The Korean decision to accelerate the HCI drive into a big push was clearly risky. Whatever its theoretical promise, an HCI big push requires the simultaneous execution of so many difficult tasks that it overstrains domestic implementation capacity and results in a problematic, lagged start-up.

Whereas Taiwan retreated in the face of entry problems with specific HCI sub-sectors such as shipbuilding and car assembly, Korea persisted until it had established a presence in the desired sector. But both countries experienced cases of policy capture by vested interests, as with auto assembly, but not on the scale of either the large low-income Asian countries or the large mid-income Latin American countries. Capture by rent-seeking groups is the Achilles' heel of industrial policy: even the East Asian escalator's mentor country, Japan, appears to have found sectoral targeting to be of dubious value (Beason and Weinstein 1994). Korean-style industrial targeting might be justified by strategic considerations such as the US withdrawal from East Asia after its defeat in Vietnam, but the economic case is weak. The new strategic trade theory has not provided the rationale for state intervention to target industry, which it had seemed likely to do in the 1980s.

The differences between Korea and Taiwan in industrial policy during their HCI drives are reflected in the diverging structure of their economies. The Korean economy is more HCI-dominated (and therefore more capital-intensive and potentially more polluting) than that of Taiwan. It is also less flexible: the Taiwanese economy is characterized by large numbers of relatively small firms which deploy relatively modest amounts of capital. The Taiwanese firms rely on skilled workers rather than capital, and they can quickly adjust to changing economic opportunities.

But neither Asian dragon attended to environmental issues until the mid-1980s, when an easing of political constraints gave the people a greater say in government. The role of democracy in economic development forms one of four themes in the concluding chapter, which reviews prospects for achieving development which is rapid, equitable, sustainable and pluralistic.

PART 7

Prospects

12

Incorporating income redistribution, environmental sustainability and democracy into economic development

Resource endowment, policy and economic growth

This chapter first summarizes the policies required to achieve rapid and equitable economic growth and then reviews the role of a country's natural resource endowment in policy choice. The basic requirements for poverty alleviation are then outlined, before the chapter examines in more depth the emerging debate over sustainable development. It concludes by stressing that effective policy implementation depends upon the formation of domestic political coalitions. Such coalitions are most appropriately analysed at a regional or national level.

Basic policy requirements

The emphasis throughout this book has been upon the overall development strategy, rather than upon more specific policy issues such as education or health care provision. This is because without rapid, equitable and sustainable growth, other policies are likely to command far fewer financial resources and to be far less effective.

The economic policies which can raise the per capita incomes of the developing countries to the levels of the industrial countries are now known (World Bank 1993a, Williamson 1993). In addition to the higher levels of investment prescribed by Rostow and others in the 1950s, it became painfully clear during the 1970s and 1980s that investment must also be efficiently allocated. The wide variation in the ability of governments to provide a climate in which investment is efficient lies behind the diverging postwar economic performance of the developing countries, summarized in Chapter 1 and analysed throughout the rest of this book.

Many governments in the less successful developing countries intervened in their economies in search of greater national autonomy (autarky) in ways which were over-ambitious and counter-productive. All too often, governments were deflected from the task of maximizing long-term welfare into the creation of rents and political patronage which could be used to maintain their political power. Meanwhile, the resulting economic distortions caused many developing countries to fall well short of their development potential, notably in sub-Saharan Africa and Latin America.

A key developmental goal which is thwarted by autarkic policies is the rapid competitive diversification of the economy in line with the country's emerging comparative advantage. Economic diversification is needed in order to reduce the risk of falling into the staple trap, whereby the fortunes of one, or at most a handful, of commodities determine economic performance. As per capita incomes rise, industrialization plays an increasingly important role in relieving the primary sector of the burden of diversification. But that diversification should initially take the form of labour-intensive manufactures. Only when the labour market turning point has been reached (and wages start to rise) should the emphasis switch towards capital- and skill-intensive manufacturing. Too many countries outside East Asia elected to leap-frog the labour-intensive stage of industrialization and attempted to move prematurely into HCI-led growth.

The success of Asian 'dragons' like Korea and Taiwan is based on two important sets of factors: favourable pre-conditions and orthodox macroeconomic policies. The first set of factors, the establishment of favourable preconditions, includes:

- the creation of a well-educated workforce (in which the early widespread provision of primary education is especially important);
- the construction of basic infrastructure in both the cities and the countryside;
- the redistribution of assets (notably land).

An important consequence of the preconditions was a widening of the opportunity for citizens to participate in the development process. This enhanced opportunity was compounded by the relatively swift move towards a labour-intensive growth path. Labour-intensive growth maximizes the number of people employed within the commercial economy and speeds the arrival of the labour market turning point.

The second important factor in the success of the East Asian countries is the pursuit of pragmatic orthodox macroeconomic policies. Such orthodox policies require the timely correction of trade and fiscal imbalances. Prompt attention to the fiscal and trade deficits kept inflation under control and provided a stable macroeconomic environment in which to invest. In addition, the priority accorded to fiscal and trade balance frequently led to the abandonment of undesirable sectoral interventions (for example, in favour of protected industries or against rural areas), before special interest groups could become entrenched and block reform.

Meanwhile, competitive diversification of the economy is fostered by the competitive exchange rate and relatively open trade policy which orthodox macroeconomic policy requires. Unlike many developing countries, the successful East Asian governments abandoned economic autarky relatively quickly in favour of intervention to encourage export diversification. But more controversially, after the turning point was reached, key HCI sectors were singled out for special assistance by Korea, and to a lesser extent Taiwan: new firms were encouraged by a package of tapered incentives to enter production.

Although such 'infant industries' matured relatively quickly in countries like Korea, it is not clear that such intervention was either beneficial or necessary. There are important examples of policy capture by rent-seeking groups in both Korea and Taiwan, although not on the scale of other large NICs.

Resource endowment and policy choice

Countries with relatively favourable natural resource endowments appear more likely to pursue lax economic policies and also to subsidize slow-maturing manufacturing sectors. Such countries are of two main types. The first group comprises large countries, including market-rich countries like India and China, as well as market-rich and resource-rich countries like Brazil and Mexico. The resource endowment of these large countries encouraged their governments to aspire to unusually high levels of self-sufficiency. The second group of countries which favoured autarkic policies comprises the small market-deficient but resource-rich primary product exporters, like the mineral economies (Fig. 12.1).

The mineral economies, in particular, mistakenly looked to the creation of a new international economic order to accelerate their growth. A key premise was the use of commodity cartels to protect primary product exporters against the believed malign role of MNCs and international capital. Two important flawed assumptions in this position were:

- that links between developing countries and international capital cannot be mutually beneficial;
- that developing countries can make effective use of large transfers of capital.

The oil-exporters show that the sudden transfer of wealth creates political pressures for over-rapid domestic spending, with adverse consequences for long-run national welfare.

In contrast, the resource-deficient Asian dragons accepted the liberal international economic order, trying to turn it to their advantage. Their small domestic markets and limited prospects for primary product exports underlined the need to abandon autarkic policies early, before rent-seeking groups could become entrenched. Foreign exchange was required with which to

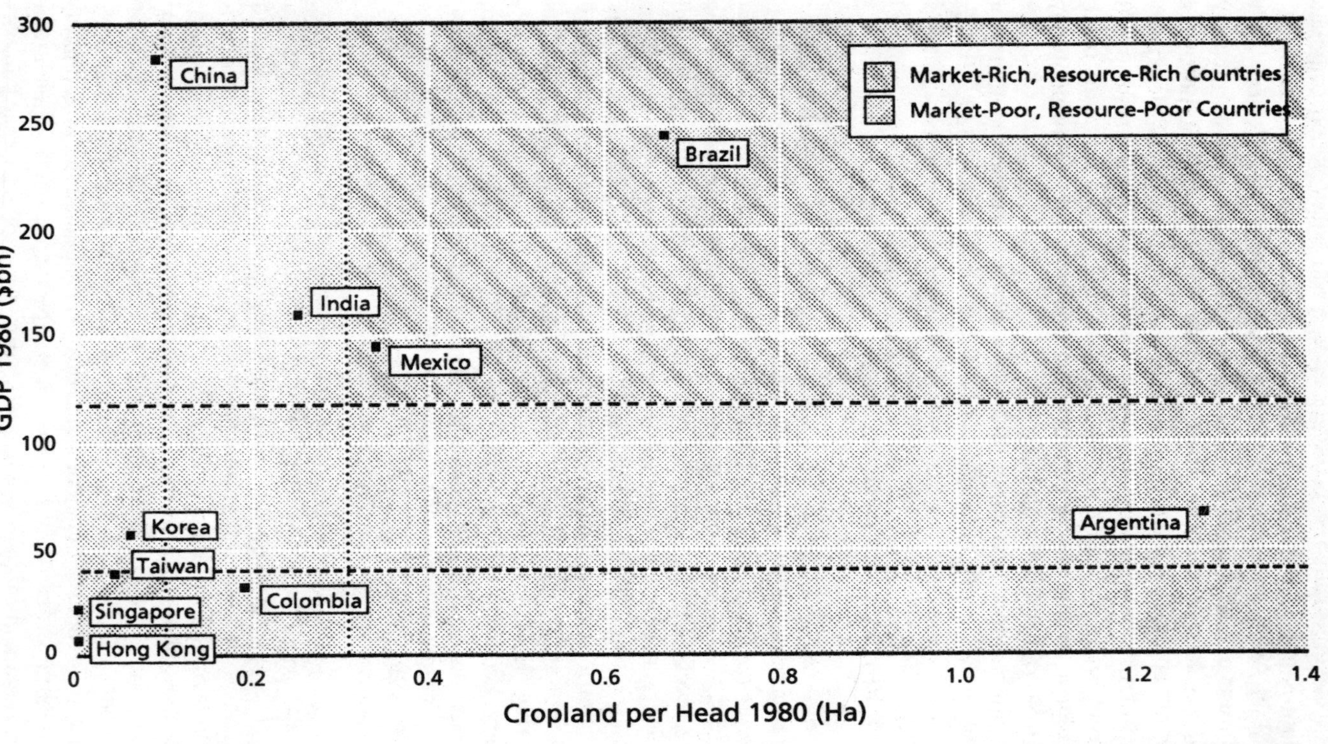

Source: WRI 1992; World Bank 1982

Fig. 12.1 *Country size/resource constraints in developing countries*

purchase key imports like oil, and this called for the early expansion of competitive labour-intensive manufacturing and the espousal of outward-oriented trade policies.

Elsewhere, the more resource-rich countries eventually found that their primary sectors (which shrank in relative importance as per capita incomes rose) could no longer support the burgeoning and inefficient protected manufacturing sector. But, by the time they realized the costs of their intervention to protect slow-maturing manufacturing, the vested interests which benefited from such policies had become sufficiently entrenched to block economic reform. In this way, a beneficial natural resource endowment may become a curse rather than a blessing.

But although the resource curse thesis is a strong recurring tendency, it is not an iron law. Like all monocausal explanations, it necessarily understates the role of other factors, such as the early East Asian stress on education and land redistribution. Moreover, there are likely to be exceptions such as resource-rich Indonesia and Malaysia, which successfully diversified first their primary sectors and then their industrial sectors. However, as Chapter 10 showed, both these resource-rich countries wavered in their commitment to prudent policies during their oil windfalls. In particular, they sought to force the pace of industrialization and created inefficient manufacturing sectors which precipitated economic crises during the mid-1980s that were potentially disastrous.

Elsewhere, the example of both India and China confirm the risks of excessively interventionist policies: they also provide clear evidence of the inefficiencies of centrally planned economies. India and China show that reform may be as difficult to achieve in command economies as in market economies, whether the command system is capitalist (India) or socialist (China). Nevertheless, by the 1980s, India and China were making strenuous efforts to reform their unusually autarkic policies, despite the fact that they had been less adversely affected by external economic shocks than most developing countries.

Reform proved problematic for the mid-income Latin American countries. The problem arose because their relatively high labour costs made adoption of the East Asian model of industrial diversification less practical for them. But the successful reforms in the more market-oriented economies of Chile and Mexico show that cultural and income differences may be weaker barriers to the diffusion of East Asian practice than has sometimes been supposed. While Chile was able to reform its incentive structure rapidly under a right-wing dictatorship, the more gradual approach favoured by Mexico may be more practical in other countries. The Mexican approach entails the rapid adoption of orthodox macro policies to stabilize the economy, and the phased withdrawal of state intervention in the hitherto protected manufacturing sectors as the economy slowly opens up to international competition.

Reform also proved problematic in sub-Saharan Africa, where Dumont (1966) was only too correct when he concluded that Africa had made a false start. An eagerness to establish modern industry and cities reflected an urban policy bias which neglected the region's comparative advantage in farming. The resulting squeeze on the farm sector and misallocation of investment in cities quickly led to slow economic growth and considerable sacrifice by the region's mainly rural population, who could ill-afford it. In much of sub-Saharan Africa, real incomes fell back to pre-1970s' levels, and renewable natural resources came under increasing pressure in rural areas. A shift towards rural development is now required with greater reliance on local initiatives rather than on central government. Malaysian-style economic diversification of, first, the primary sector and then, competitive labour-intensive manufacturing is needed.

Income redistribution

The economic policies which will allow developing countries to achieve steady economic growth of 6–7 per cent annually are known. It is also clear which policies promote more egalitarian income distribution, but the domestic political obstacles to the achievement of both goals remain. Governments that adhered to more autarkic industrial policies postponed the labour market 'turning point'. Consequently, a relatively small but privileged worker aristocracy enjoyed high wages in the protected and uncompetitive manufacturing sector, along with those in the professions. The majority of workers remained trapped in low-income low-productivity 'marginalized' jobs of artisan manufacturing and petty services.

The East Asian model shows that a relatively equitable income distribution can be sustained throughout the rise from low-income to high-income status. Although the labour-intensive phase of that model, based on light industry exports, entailed working long hours and low wages, surplus labour was rapidly absorbed into the economy. Rural workers flocked to cities where there were more family job opportunities and, moreover, where they experienced substantial real income gains, once the turning point had been reached. In fact, Taiwan managed to achieve a steady decline in income inequality as its average per capita income rose. Table 4.4 shows that for Taiwan and Korea, the ratio of the income accruing to the richest quintile compared with the poorest quintile has been around 4 and 7, respectively. This compares favourably with the developed countries, and is far superior to Latin America and also to sub-Saharan African countries like the Ivory Coast and Kenya.

Griffin (1989) identifies five key requirements for the achievement of a more egalitarian pattern of growth, namely:

- investment in human capital (i.e. education);
- early redistribution of key productive assets (notably land);
- pursuit of an employment-intensive development policy;

- sustained rapid growth in aggregate per capita income;
- encouragement of local participation in the choice and implementation of social and economic projects.

With the possible exception of the latter element, Griffin might have been describing the East Asian development model, with its stress on orthodox macroeconomic policies and export-led economic diversification.

The adoption of such policies within individual countries, however, poses problems. For example, the successful industrialization of Taiwan and Korea drew heavily upon early land reforms which transferred assets to small peasant farmers. Attempts to redistribute assets have proved less promising elsewhere, as countries as diverse as Brazil, the Philippines, Sri Lanka, El Salvador and India show. But simulations of redistributive strategies by Chenery *et al.* (1974) suggest that land reform may be less crucial than some suppose. Chenery compared the welfare implications of redistributive policies (which included wage restraint, direct consumption transfers and direct investment transfers) with a base case growth-oriented strategy.

In the simulations, neither wage restraint nor consumption transfers improved the welfare of the poor, compared with the base case growth-oriented scenario. Moreover, although redistribution of assets did lead to some modest improvement in the welfare of the poor, the practicality of such a strategy is clearly open to doubt. This is because it is unlikely that political realities would permit any government to sustain the level of transfers from rich to poor that are required. Such transfers were assumed by Chenery to be equivalent to 2 per cent of GNP annually, and to take the form of infrastructure improvements for the poor.

It is therefore fortunate that, as shown by the East Asian development model, the same policies which raise the rate of economic growth and curb the wasteful use of resources (and especially rural resources) are also likely to substantially assist the least well off. Consequently, contrary to the assumptions of earlier theorists like Lewis (1954) and Williamson (1965), a substantial degree of income inequality may not be the inevitable price of economic growth. In fact, income inequality is more likely to result from the pursuit of strongly interventionist policies, notably autarkic industrialization.

Brazil most clearly demonstrates that autarkic policies establish rent-seeking groups which benefit at the expense of weaker groups in society. Brazil also shows that, ironically, such rent-seeking groups oppose reform, on the grounds that the deflation associated with orthodox economic adjustment will hit the poor hardest. What is required in such a case is a more labour-intensive development strategy that expands the opportunity for hitherto marginalized workers to participate in the development process. If that is achieved, assistance in the form of government transfers of resources can be targeted on the poor who remain excluded from the development process. Such assistance should preferably take the form of subsidized health care and primary education (Squire 1993). But, once again, Brazil over the past 3 decades underlines the political obstacles to policy reform. Meanwhile, those countries which have successfully crossed that hurdle, like Chile and Mexico, still need to achieve sustainable development.

Sustainable economic growth

It is now widely accepted that it is necessary to ensure that economic growth does not occur at the expense of a long-term deterioration in the natural environment and, therefore, at the expense of the welfare of future generations. In the words of the Brundtland Report (World Commission on Environment and Development 1987), what is needed is 'development that meets the needs of the present [generation] without compromising the ability of future generations to meet their own needs.'

Three categories of natural resources may be recognized which comprise renewable, non-renewable and non-renewable global resources (Mikesell 1992). Renewable resources like forests, soils and fisheries need to be managed, so that stocks are not exhausted. Indeed, such resources have the potential to be expanded, provided depletion does not cross a critical ecological threshold. The costs and benefits of alternative levels of both production and conservation can be readily calculated for such resources and can then be incorporated into economic transactions and markets by appropriate price adjustments. There is clear evidence that the amelioration of environmental damage through the incorporation of such price adjustments can be most effectively achieved under conditions of rapid and equitable economic growth (World Bank 1992a).

The non-renewable fund resources, like minerals and fossil fuels, present more of a problem for sustainable development, because their use implies a once-for-all loss of an environmental asset. In this case, in order to ensure that asset depletion is not at the expense of future generations, it is necessary that a sufficient amount is invested annually out of the net revenues (total revenue minus the cost of labour and capital). This investment will provide an alternative source of income for the depleted resource when it is exhausted, or a substitute resource service (for example, using technology to substitute a new material for depleted copper, or solar energy to replace depleted oil reserves).

The greatest environmental management challenge is posed, however, by the non-renewable global resources (atmosphere, oceans, biodiversity) which are essential to life and whose degradation may be irreversible. This is because the costs and benefits involved in the use of such resources are difficult to calculate, so that their management rests strongly on value judgements. Table 12.1 shows the spectrum of perspectives on environmental issues and sustainable development. The spectrum may be summarized in terms of two basic categories: the strong green and weak green approaches.

The strong green environmental approach

The strong green stance likens natural resources to a capital stock which the present generation must not consume. The strong green approach, in effect, places an absolute value on the environment, and in more extreme cases (O'Riordan 1988) may advocate that the environment possesses the right to existence, irrespective of human welfare considerations (Table 12.1). Others advocate that special rules should be applied to environmental resources. In particular, the conventional use of discount rates to compare present costs and benefits with future costs and benefits should be suspended as far as environmental issues are concerned. This is because such discounting tends to place a very low value on future outcomes. Yet the distant future is exactly when environmental damage is likely to have cumulated to high levels. In effect conventional discounting procedures tend to postpone responses to long-term problems.

A simple example will show how the discounting procedure down-weights the importance of costs and benefits that are incurred far into the future. Using a conventional discount rate of 8 per cent, 1 pound in 10 years' time is worth only 43 pence in present value terms. One pound in 100 years' time is worth only 0.05 pence in present value terms. These examples show how discounting places a relatively low value on future benefits and costs, including environmental costs which if neglected may inflict irreversible damage.

Global warming provides a classic example of the consequences of using conventional discount rates. Cline (1992) puts the strong green approach by arguing that a special low discount rate of just under 2 per cent should be applied to the costs and benefits of reducing the carbon dioxide emissions associated with global warming. The 2 per cent rate is the social rate of time preference, whereas a rate of 8 per cent reflects the private opportunity cost of capital. The lower social rate reflects the greater ease with which society can bear individual project investment risks, whereas the private discount rate tends to be higher to allow for project risks and the taxation of project income.

Cline estimates that the costs of adjustment to environmental damage (equivalent to 3.5 per cent of GDP annually) will be incurred abruptly within the next generation (Fig. 12.2). The benefits (estimated at up to 16 per cent of GDP annually under the 'high-damage scenario') will largely accrue, however, several generations from the present. Although the ***undiscounted*** benefits exceed the costs, conventional discounting rules (based on the opportunity cost of capital, i.e. the returns on competing investment options) would not sanction such an expenditure. This is because the discounted costs, being closer to the present, are much higher than the discounted benefits which accrue far into the future and are therefore subjected to much greater discounting. More specifically, each pound of costs incurred in 20 years to avoid global warming would be discounted at 8 per cent to 17 pence, whereas each pound of benefits gained from such action and accruing, say, one hundred years into the future, has a value of only 0.05 pence (less than one-three-hundredth as much).

Table 12.1 *Perspectives on sustainable development*

Technocentric (overlapping categories) Cornucopian	Accommodating	Ecocentric Communalist	Deep ecology
Green labels			
Resource exploitative position	Resource conservationist and 'managerial' position	Resource preservationist	Extreme preservationist position
Type of economy			
Anti-green economy, unfettered free markets	Green economy, green markets guided by economic incentive instruments (Els) (e.g. pollution charges, etc.)	Deep green economy, steady-state economy regulated by macroenvironmental standards and supplemented by Els	Very deep green economy, heavily regulated to minimize 'resource-take'
Management strategies			
Primary economic policy objective, maximize economic growth (max. GNP)	Modified economic growth (adjusted green accounting to measure GNP)	Zero economic growth; zero population growth	Reduced scale of economy and and population
Taken as axiomatic that unfettered free markets in conjunc-	Decoupling important but infinite substitution rejected. Sustainability	Decoupling plus no increase in scale. 'Systems' perspective - 'health' of	Scale reduction imperative; at the extreme for some there is a literal interpretation of Gaia

tion with technical progress will ensure infinite substitution possibilities capable of mitigating all 'scarcity/limits' constraints (environmental sources and sinks)	rules: constant capital rule. Therefore some scale changes	whole ecosystem very important; Gaia hypothesis and implications	as a personalized agent to which moral obligations are owed
Ethics			
Support for traditional ethical reasoning: rights and interests of contemporary individual humans; instrumental value (i.e. of recognized value to humans) in nature	Extension of ethical reasoning: 'caring for other' motive - intragenerational and intergenerational equity (i.e. contemporary poor and future people)	Further extension of ethical reasoning: interests of the collective take precedence over those of the individual; primary value of ecosystems and secondary value of component functions and services	Acceptance of bioethics (i.e. moral rights/interests conferred on all non-human species and even the abiotic parts of the environment); intrinsic value in nature (i.e. valuable in its own right regardless of human experience)
Sustainability labels			
Very weak sustainability	Weak sustainability	Strong sustainability	Very strong sustainability

Source: Turner *et al.* (1994)

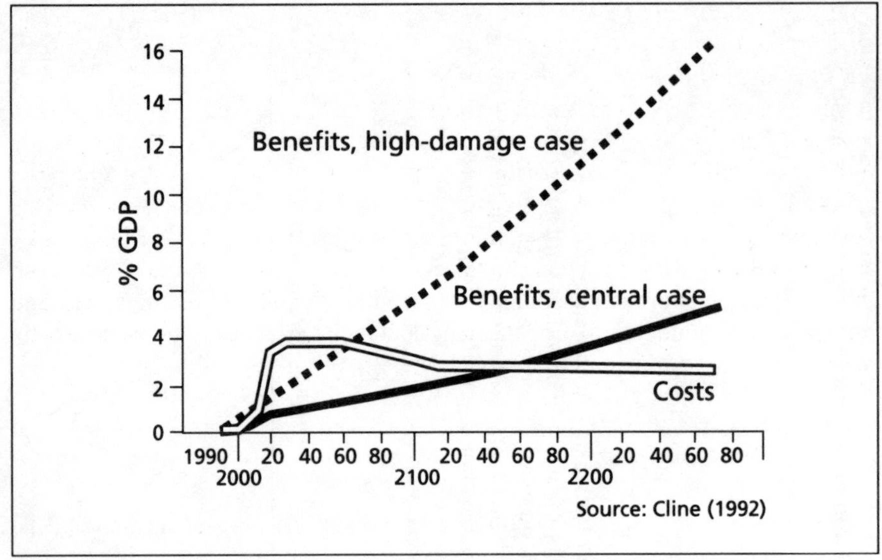

Fig. 12.2 *Costs and benefits of strong action to abate greenhouse warming*

The strong green stance also tends to stress a concern for intra-generational equity (that is, equality of income now). This leads to calls for radical changes in developed country lifestyles in order to reduce their claims on natural resource consumption. This is because the rich are perceived to benefit at the expense of the poor in the *present* generation, as well as at the expense of future generations. The exponents of this view argue in support of the large-scale transfer of capital and technical resources to the developing countries. They claim that such large transfers will prevent further erosion of the physical resource base – notably soil, forest and water resources. But much of the evidence presented in this book suggests that such a solution ignores the political realities of many developing countries, whose governments would be hard-pressed to make effective use of such transfers.

The weak green environmental approach

The alternative weak green stance is concerned with the maximization of human welfare through the rational comparison of the social costs and benefits of alternative resource use options. Peskin (1993) argues that this is a more humanitarian approach to environmental issues than the strong green approach, because it places a higher value on human welfare. The weak green approach builds on orthodox development economics by seeking to add in environmental costs and benefits, rather than by singling out the environment for special treatment as the overriding priority.

For example, the weak green approach considers that future generations may be helped as much by depleting a resource now and using the proceeds to educate children, as by leaving the resource untouched; or by using the proceeds from forest exploitation to invest in infrastructure, rather than by pursuing more restrained exploitation; or by expanding scientific knowledge as opposed to curbing carbon dioxide emissions (Summers 1992). In effect, the weak green approach assumes that man-made capital (building stock) and human capital (education) can in many instances substitute for environmental capital. The strong green approach vehemently rejects this view.

The weak green position opposes the application of a different set of discount rates for the environment. Rather, the key test for an investment is the net present value. In this view, those projects should be selected which give a higher return than the next best alternative use of scarce investment resources. This implies that projects are selected which maximize both social welfare and the return on capital. But this will only be the case if the costs of the investment include all social costs, including the cost of environmental damage.

The weak green approach rejects the strong green assumption that higher welfare now, based on the consumption of natural resources, is at the expense of future generations. The reason for this is that resources which are exploited now will raise welfare. The more effectively capital is used, the faster social welfare improvements can occur and the wealthier the population the greater the willingness and ability to combat environmental damage.

Efficient capital use calls for investment in those projects which yield the highest net present value after the costs and benefits have both been discounted into present value terms. For example, if a low (2 per cent) discount rate is used, a pound invested will yield only £52.50 in real terms over 200 years, whereas at a higher rate (8 per cent) it will yield £4 839 100. The adoption of a low rate of discount for environmental projects will therefore tend to depress the productivity of capital.

In the developing countries, the higher living standards which accrue from the more efficient use of capital will slow population growth and limit the environmental damage. Such environmental damage is aggravated by the short time horizons associated with extreme poverty. The very poorest farmers tend to exhibit very high rates of discount which, for example, are of the order of 30 to 40 per cent for investments in reducing soil erosion in India. This means that such farmers demand a very high return from an investment before they will accept the sacrifice it entails in terms of forgoing consumption now.

In the case of the industrial countries, Summers (1992) points out that the retention of conventional discounting in environmental matters has led to the more affluent nations being better able to control pollution. For example,

West Europeans in the 1990s were 7 times better off than their forebears were 60 years earlier, because of the effective use of capital. Similarly, future generations will also benefit from high discount rates because rapid economic growth will endow them with much greater financial resources and also with associated technical skills that will help resolve a wide range of problems, including environmental problems.

But the weak green approach needs to take note of the risk that environmental damage may prove to be irreversible. This can be incorporated into the analysis by giving a higher weighting to investment in environmental benefits (such as conservation), by making special allowance for the value options (i.e. future opportunities) which they preserve. Such modifications will tend to increase the costs of environmental damage and, therefore, to increase the benefits which flow from investment to curb that damage.

At the core of the weak green argument is the view that rigid rules and special discount rates should not substitute for the careful incorporation of environmental costs into investment decisions. Provided that the costs and benefits of alternative options are properly valued and carefully considered, it is in the social interest to undertake investments which yield the best (highest) economic return.

Three sets of environmental policy options

The weak green view tends to accord lower priority to long-term problems (biodiversity and global warming) than to medium-term ones (soil erosion, contaminated water, cleaner air). It argues that policies which target medium-term problems are likely to benefit the poorest most, whereas the richer inhabitants of the developing countries are in a position to protect themselves from waterborne disease and air pollution (by, for example, selecting the safest places to live). Similarly, the industrial countries have been able to use their wealth to substantially reduce pollutants since 1970 and to extend conservation.

More to the point, given prevailing political realities, the weak green approach has a better chance of being implemented than the more radical strong green agenda. With this in mind, the World Bank (1992a) has identified three sets of environmental policy, some of which would be very inexpensive to adopt (Fig. 12.3). First, there are policies which involve no financial costs, but which carry both economic and environmental benefits. Examples include the removal of subsidies on energy, fertilizer, pesticides, irrigation water and ranching, which merely serve to encourage the wasteful use of resources. Similarly, taxes on urban road congestion (which, in effect, impose upon motorists the costs of noise and pollution which they inflict on others) are also included in this category.

A second set of environmental policies involve making expenditures but have positive net economic benefits, even without the inclusion of environmental gains. These include many policies which help the poorest in society,

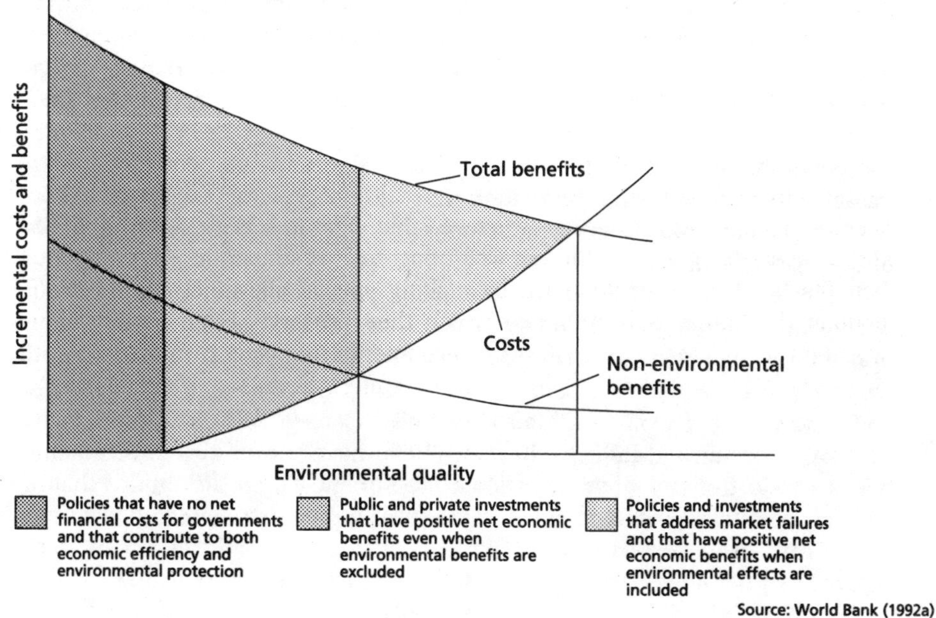

Fig. 12.3 *Benefits and costs of environmental policies*

such as improved water supplies and soil conservation. Also included in this category are policies which increase investment in female education and thereby contribute to a significant reduction in the birth rate, as well as to gains in nutrition, hygiene and health.

Finally, there are those policies which entail public investment to deal with market failure (that is, markets where critical costs such as environmental damage have been neglected) and which have net positive economic benefits when the environmental effects are included. They include pollution taxes and the regulation of hazardous waste disposal. But consistent with the welfare-maximizing goal of the weak green approach, the pursuit of such policies proceeds only as long as the net social benefits outweigh the net social costs (Fig. 12.3). But none of these three sets of policies faces much chance of implementation in a specific country if the political consensus required to carry them out is lacking.

Political economy

Governments will need to intervene in the economy not only to provide stable macroeconomic conditions and infrastructure, but also in order to steer the economy towards more equitable and less environmentally degrading growth. Such intervention, however, carries the risk of a return to the damaging rent-dispensing systems of the 1970s. To avoid such a relapse will require imaginative local, national and global solutions, guided by the principle that state intervention should seek to correct market failures (whether to provide universal primary education or to incorporate natural resource depletion into national accounts). Such intervention must stress greater equality of opportunity and concern for future generations – but without sacrificing the benefits of the market's allocative efficiency.

The role of the state in most developing countries increased too rapidly during the postwar period. Whereas the share of the public sector in non-agricultural employment averaged only 24 per cent in the developed countries, it expanded to an average 44 per cent in the developing countries, and topped 70 per cent in the states which espoused socialism like Ghana, Zambia, Tanzania and India. Gelb *et al.* (1986) have shown how governments may corrode the efficiency of national investment within little more than a decade. Unions in protected industrial sectors and ministries bid up wages to 5 or 10 times those of rural areas. This encourages a shift of labour from potentially productive farming and into unproductive urban activities.

With hindsight, it is clear that much postwar development theory rather naively assumed that the role of the government was the maximization of long-term social welfare. Such naive assumptions had their antithesis in the cynical analyses of the dependency theorists concerning the behaviour of political elites. In this context, the new political economy associated with Bates (1981) makes the more realistic assumption that governments arbitrate between competing interest groups and provide rewards in exchange for political support. This view makes it easier to understand why government intervention can so easily degenerate into a system for dispensing rents, whether to party officials in the one-party state or to clientelistic groups within Latin American political systems.

Nevertheless, this book shows that it is possible to be optimistic about constraining the political risk associated with more active state intervention. This optimism is based on the fact that sufficient experience has now been accumulated concerning the process of development to show which policies can solve the economic, social and environmental problems created by the desire to raise human welfare world-wide. That experience also shows how costly policy failure can be. Political systems must be devised which mediate between competing domestic interest groups in such a way as to prevent policy from being captured by narrow interest groups. If the policy convergence implied by the 'Washington Consensus' *does* hold, and dramatic changes in domestic economic policy are thereby avoided, democratic political systems are most desirable. This is because democratic governments are likely to be most sensitive to social and environmental goals.

But the political coalitions needed to sustain desirable policies must be tailored to the specific conditions of individual countries. This is because, unlike the economic and environmental issues, they are strongly influenced by local idiosyncrasies and have so far proved much too complex to be amenable to 'general' answers. Consequently, this book will have performed a useful function if it provides a stepping stone to the more detailed analysis of the political economy of development which must be undertaken at a regional or a national level.

References

ACHARYA, S.N. 1981: Perspectives and problems of development in sub-Saharan Africa. *World Development* 9, 109–47.

ACHEBE, C. 1958: *Things Fall Apart*. London: Heinemann.

ADAMSON, A.H. 1972: *Sugar Without Slaves: The Political Economy of British Guiana 1838–1904*. New Haven CT: Yale University Press.

AHLUWALIA, I.J. 1985: *Industrial Growth In India: Stagnation Since the Mid-1960s*. New Delhi: Oxford University Press.

AHMAD, Y.J., EL SERAFY, S. and LUTZ, E., 1989: *Environmental Accounting for Sustainable Development*. Washington DC: World Bank.

AIKEN, S.R. 1981: Squatters and squatter settlements in Kuala Lumpur. *Geographical Review* 71, 158–75.

ALONSO, W. 1971: The economics of urban size. *Proceedings of the Regional Science Association*.6, 149–57.

AMSDEN, A. 1985: The state and Taiwan's economic development. In: Evans, P.B. (ed.) *Bringing the State Back In*. Cambridge: Cambridge University Press, 78–106.

AMSDEN, A 1989: *Asia's Next Giant: South Korea and Late Industrialization*. New York: Oxford University Press.

ANDRAIN, C.F. 1988: *Political Change in the Third World*. Boston MA: Unwin Hyman.

ARMSTRONG, H. 1993: Regional inequalities and regional policy in Great Britain. *Developments in Economics* 9, 51–70.

AUTY, R.M. 1975: Scale economies and plant vintage: Toward a factory classification. *Economic Geography* 51, 150–62.

AUTY, R.M. 1976: Caribbean sugar factory size and survival. *Annals of the Association of American Geographers* 66, 76–88.

AUTY, R.M. 1986: Multinational resource corporations, nationalization and diminished viability. In: Dixon, C.J., Drakakis, Smith, D. and Watts, H.D.(eds) *Multinational Corporations and the Third World*. London: Croom Helm, 160–87.

AUTY, R.M. 1987: The product life-cycle and the global location of energy-intensive industry after the second oil shock. In: Muegge, H., Stohr, W.B., Hesp, P. and Stuckey. B. (eds). *International Economic Restructuring and the Regional Community*. Aldershot: Avebury, 92–115.

AUTY, R.M. 1990a: *Resource-Based Industrialization: Sowing the Oil in Eight Exporting Countries*. Oxford: Clarendon Press.

AUTY, R.M. 1990b: The impact of heavy industry growth poles on South Korean spatial structure. *Geoforum* 21, 23–33.

AUTY, R.M. 1992: The macro impact of Korea's HCI drive re-evaluated. *Journal of Development Studies* 28, 24–48.

AUTY, R.M. 1993: Multinational resource corporations, the product life cycle and product strategy: the oil majors' response to heightened risk. *Geoforum* 14, 1–13.

AUTY, R.M. 1993: *Sustaining Development in Mineral Economies: The Resource Curse Thesis*, London: Routledge.

AUTY, R.M. 1994: Industrial policy reform in six large NICs: The resource curse thesis. *World Development* 22, 11–26.

AUTY, R.M. and WARHURST, A. 1993: Sustainable development in mineral exporting countries. *Resources Policy* 19, 14–29.

BACHA, E.L. 1986: External shocks and growth prospects: case of Brazil 1973–79. *World Development* 14, 919–36.

BAER, W. 1989: *The Brazilian Economy*. New York: Praeger.

BALASSA, B. 1982: Structural adjustment policies in developing countries. *World Development* 10, 23–38.

BALASSA, B. 1985: Adjusting to external shocks: the newly industrializing developing economies in 1974–76 and 1979–81. *Weltwitschaftsliches Archiv* 12, 116–41.

BALASSA, B. 1988: The lessons of East Asian development: An overview. *Economic Development and Cultural Change* 36, Supplement, 273–90.

BALASUBRAMANYAM, V.N. 1984: *The Economy of India*. London: Weidenfeld and Nicholson.

BALDWIN, R.E. 1956: Patterns of development in newly settled regions. *Manchester School of Social and Economic Studies*, 24, 161–79.

BARKIN, D. 1975: Regional development and interregional equity: A Mexican case study. In: Cornelius, W.A. and Trueblood, F.M. (eds) *Urbanization and Inequality: Political Economy of Urban and Rural Development in Latin America*. London: Sage, 277–99.

BARNETT, D.F. and SCHORSCH, L. 1983: *Steel; Upheaval in a Basic Industry*. Cambridge MA: Ballinger Publishing.

BATES, R.H. 1981: *Markets and States in Tropical Africa*. Berkeley CA: University of California Press.

BEASON, R. and WEINSTEIN, D. 1994: Growth, economies of scale, and targeting in Japan (1950–90). *Discussion Paper* 1644. Cambridge MA: Harvard Institute of Economic Research.

BEAUMONT, P., BLAKE, G.H. and WAGSTAFFE, J.M. 1988: *The Middle East: A Geographical Study*. London: David Fulton.

BECKFORD, G. 1972: *Persistent Poverty*. Oxford: Oxford University Press.

BEENSTOCK, M. 1983: *The World Economy in Transition*. London: George Allen and Unwin.

BEIER, G., CHURCHILL, A., COHEN, M. and RENAUD, B. 1976: The task ahead for the cities of the developing countries. *World Development*, 4, 363–409.

BELL, M. and ROSS-LARSON, B. 1984: Assessing the performance of infant industries. *Journal of Development Economics*, 16, 101–28.

BELLO, W. and ROSENFELD, S. 1990: *Dragons in Distress*. London: Penguin.

BERGMAN, E.F. 1971: The multinational corporation. *Professional Geographer* 23, 255–60.

BERGSMAN, J. 1970: *Brazil: Industrialization and Trade Policies*. Oxford: Oxford University Press.

BERGSTEN, C.F. and NOLAND, M. 1993: *Reconcilable Differences? US–Japan Economic Conflict*. Washington DC: IIE.

BERISTAIN, J. and TRIGUEROS, I. 1990: In: Williamson, J. (ed.) *Latin American Adjustment*. Institute for International Economics: Washington DC, 154–77.

BERNSTAM, M.S. 1991: *The Wealth of Nations and the Environment*. London: Institute of Economic Affairs.

BERRY, J.B.L. and GARRISON, W.L. 1958: Alternative explanations of urban rank-size relationships. *Annals of the Association of American Geographers* 48, 83–91.

BHAGWATI, J. 1993: *India in Transition: Freeing the Economy*. Oxford: Clarendon Press.

BHARGAVA, P.K. 1987: India: Can we eliminate poverty? *Long Range Planning*, 20, no. 2, 21–31.

BHATTACHARYA, A. and PANGESTU, M. 1993: *Indonesia: Development Transformation Public Policy*. Washington DC: World Bank.

BINSWANGER, H.P. 1991: Brazilian policies that encourage deforestation in the Amazon. *World Development* 19, 821–30.

BINSWANGER, H. and PINGALI, P. 1988: Technological priorities for farming in Sub-Saharan Africa. *World Bank Research Observer* 3, 81–98.
BIRDSALL, N. and STEER, A. 1993: Act now on global warming – but don't cook the books. *Finance and Development* 30, 6–8.
BOSE, N.K. 1968: Calcutta: A premature metropolis. In: *Cities*. New York: Scientific American, 58–74.
BOSERUP, E. 1965: *The Conditions of Agricultural Growth*. London: Allen and Unwin (reprinted in 1993 in London by Earthscan).
BOVENTER, E.G. von 1975: Regional growth theory. *Urban Studies* 12, 1–29.
BRANDER, J. and SPENCER, B. 1985: Export subsidies and market share rivalry. *Journal of International Economics* 18, 83–100.
BRANDON, C. and RAMANKUTTY, R. 1993: *Toward an Environmental Strategy for Asia*. Washington DC: World Bank.
BREESE, G. 1966: *Urbanization in Newly Developing Countries*. New York: Prentice Hall.
BROWNING, D. 1971: *El Salvador: Landscape and Society*. Oxford: Oxford University Press.
BRUTON, H. 1989: Import substitution. In: Chenery, H. and Srinivasan, T. (eds) *Handbook of Development Economics*. Amsterdam: North Holland, 1602–44.
BUNGE, F.M. 1981: *China: A Country Study*. Washington DC: America University.
BUTTERWORTH, D. and CHANCE, J.K. 1981: *Latin American Urbanization*. Cambridge: Cambridge University Press.
CARAVAL, M.J. and GEITHMAN, D.T. 1974: An economic analysis of migration in Costa Rica. *Economic Development and Cultural Change* 23, 105–22.
CEPD 1991: *Taiwan Statistical Yearbook 1991*. Taipei: Council for Economic Planning and Development.
CHAKRAVARTI, A.K. 1973: The green revolution in India. *Annals of the Association of American Geographers* 63, 319–30.
CHANG, J. 1991: *Wild Swans*. London: HarperCollins.
CHAU, L.C. 1993: *Hong Kong: A Unique Case of Development*. Washington DC: World Bank.
CHEN, X. 1991: China's city hierarchy, urban policy and spatial development in the 1990s. *Urban Studies* 28, 341–67.
CHENERY, H.B. 1981: Restructuring the world economy. *Foreign Affairs* 59, 1102–20.
CHENERY, H., AHLUWALIA, M.S., BELL, C.L.G., DULOY, J.H. and JOLLY, R. 1974: *Redistribution with Growth*. Oxford: Oxford University Press.
CHENERY, H.B., ROBINSON, S. and SYRQUIN, M. 1986: *Industrialization and Growth*. New York: Oxford University Press.
CHERU, F. 1992: Structural adjustment, primary resource trade and sustainable development in sub-Saharan Africa. *World Development* 20, 497–512.
CHILCOTE, R.H. 1974: Dependency: A critical synthesis of the literature. *Latin American Perspectives* 1, 4–29.
CHO, W-D. 1989: Scale economies and entry regulation: development of the Korean auto industry. Unpublished D Phil Dissertation, Oxford University.
CLAYTON, E.S. 1964: *Agrarian Development in Peasant Economies*. Oxford: Pergamon Press.
CLINE, W. 1992: *The Economics of Global Warming*. Washington DC: Institute for International Economics.
COALE, A.J. 1964: The population dilemma. In: Novak, T. and Lekachman, D. (eds) *Economy and Society*. New York: Columbia University Press, 125–40.
COHEN, A. and GUNTER, F.R. 1992: *The Colombian Economy*. Boulder CO: Westview Press.
COLE, J.P. 1987: Regional inequalities in the People's Republic of China. *TESG* 78, 201–13.

COLE, J.P. 1988: China's aspirations and reality. Working Paper, Department of Geography, Nottingham University.
CORBO, V. and de MELO, J. 1987: Lessons from the southern cone policy reforms. *World Bank Research Observer* 2, 111–42.
CORBO, V. and NAM, S.W. 1988: Korea's macroeconomic prospects and policy issues for the next decade. *World Development* 16, 35–45.
CORDEN, W.M. 1982: Booming sector and Dutch disease economics: A survey. *Australian National University Working Paper* 78, Canberra: Australian National University.
CRANDALL, R.W. 1981: *The US Steel Industry In Recurrent Crisis*. Washington DC: Brookings Institution.
CROOK, C. 1990: World Trade. *The Economist*, September 22, centre supplement.
CROOK, C. 1991: Caged: A survey of India. *The Economist*, May 4, centre supplement.
DAVIS, K. 1965: The urbanization of the human population. *Scientific American* 213, Sept., 40–53.
de SA, P. and MARQUES, I. 1985: The Carajas iron ore project. *Resources Policy* 11, 245–56.
DESHPANDE, L.K. 1986: The Seventh Plan and some aspects of employment. *Indian Economic Journal* 33, 160–7.
DHOLAKIA, R.H. 1986: Sources of economic growth in India implied in the Seventh Year Plan 1985–90. *The Indian Economic Journal* 33, 160–7.
DICKINSON, J.C. 1972: Alternatives to monoculture in the humid tropics. *Professional Geographer* 24, 217–22.
DINWIDDY, B. 1973: *Current and Future Trends in Development Assistance*. London: ODI.
DIXIT, A.K. 1988: Optimal trade and industry policies for the US automobile industry. In: Feenstra, R. (ed.) *Empirical Methods for International Trade*. Cambridge MA: MIT Press, 141–65.
DORNBUSCH, R. and EDWARDS, S. 1991: *The Macroeconomic Problems of Populism*. Chicago IL: University of Chicago Press.
DORNBUSCH, R. and PARK, Y.C. 1987: Korean growth policy. *Brookings Papers on Economic Activity*, No. 2, 389–454.
DOS SANTOS, T. 1969: The crisis of development theory and the problem of dependence in Latin America. In: Bernstein, H. (ed.) *Underdevelopment and Development*. London: Penguin, 57–80.
DUBOIS, V.D. 1974: *The Drought in West Africa: Evolution, Causes and Physical Consequences*. American Universities Fieldstaff Reports, West Africa Series, 15, no. 1.
DUMONT, R. 1966: *False Start in Africa*. London: André Deutsch.
DUMONT, R. and MOTTIN, M-F. 1983: *Stranglehold in Africa*. London: André Deutsch.
DUNCAN, R.C. 1993: Agricultural export prospects for sub-Saharan Africa. *Development Policy Review* 11, 31–45.
DUNNING, J.H 1993: *Multinational Enterprises and the Global Economy*. London: Addison-Wesley.
DURHAM, W.H. 1979: *Scarcity and Survival in Central America*. Stanford CA: Stanford University Press.
ECKSTEIN, S. 1977: *The Poverty of Revolution*. Cambridge: Cambridge University Press.
ECKSTEIN, S. 1990: Urbanization revisited: Inner city slum of hope and squatter settlement of despair. *World Development*, 18, 165–81.
ECONOMIST, THE 1983: Drought in Africa. September 10, 45–6.
ECONOMIST, THE 1986: Grass saves India's soil. September 20, 97–8.
ECONOMIST, THE 1987: China beyond the wall. September 12, 21–4.

ECONOMIST, THE 1988: Come back multinationals. November 26, p. 103.
ECONOMIST, THE 1991: Smog City. May 18, p. 80.
ECONOMIST, THE 1992: Salt of the earth. June 27, p. 84.
ECONOMIST, THE 1993a: Demography. February 20, p. 137.
ECONOMIST, THE 1993b: Everybody's favourite monsters: a survey of multinationals. March 27, centre supplement.
EDEL, M. 1972: Planning, market or warfare? Recent land use conflict in American cities. In: Edel, M. and Rothenberg, J. (eds) *Readings in Urban Economics*. London: Collier Macmillan, 134–51.
EDEN, M.J. 1978: Ecology and land development: The case of the Amazon rainforest. *Transactions of the IBG*, 3, 444–63.
ELKAN, W. 1973: *Introduction to Development Economics*. London: Penguin.
ELKAN, W. 1978: *An Introduction to Development Economics*. Harmondsworth: Penguin.
ENOS, J. L. and PARK, W.H. 1988: *The Adoption and Diffusion of Imported Technology: The Case of Korea*. London: Croom Helm.
FAO 1986: *Ethiopia: Highlands Reclamation Study: Final Report*. AG:UTF/ETH/037, Rome: FAO.
FARMER, B.F. 1983: *An Introduction to South Asia*. London: Methuen.
FINANCIAL TIMES 1992: Political resentment thrives in a country divided. December 29.
FLATTERS, F. and JENKINS, G. 1986: Trade policy in Indonesia. Mimeo, Cambridge MA: HIID.
FOX, D.J. 1983: Central America, including Panama. In: Blakemore, H. and Smith, C.T. (eds) *Latin America: Geographical Perspectives*. Oxford: Oxford University Press, 133–85.
FRIEDMANN, J. 1966: *Regional Development Policy: A Case Study of Venezuela*. Cambridge MA: MIT Press.
FRIEDMANN, J. 1973: *Urbanization, Planning and National Development*. London: Sage.
GEERTZ, C. 1963: *Agricultural Involution: The Process of Ecological Change in Indonesia*. Berkeley CA: University of California Press.
GELB, A. 1988: *Oil Windfalls: Blessing or Curse?*. New York: Oxford University Press.
GELB, A.H., KNIGHT, J.B. and SABOT, R.H. 1986: Lewis through the looking-glass: public sector employment, rent-seeking and economic growth. Mimeo, Cambridge MA: Employment and Enterprise Policy Analysis, Havard Institute for International Development.
GILBERT, A. 1994: *The Latin American City*. London: Latin American Bureau.
GILBERT, A. and GUGLER, J. 1992: *Cities, Poverty and Development: Urbanization in the Third World*. Oxford: Oxford University Press.
GILLIS, M. 1984: Episodes in Indonesian economic growth. In: Harberger, A.C. (ed.) *World Economic Growth*. Stanford CA: ICS Press, 31–64.
GIRVAN, N. 1971: *Foreign Capital and Economic Underdevelopment in Jamaica*. Woking: Unwin Brothers.
GOLDMAN, A. 1993: Agricultural innovation in three areas of Kenya: Neo-Boserupian theories and regional characterization. *Economic Geography* 69, 44–71.
GOLDSMITH, W.W. and WILSON, R. 1991: Poverty and distorted industrialization in the Brazilian northeast. *World Development* 19, 435–55.
GOOD, K. 1985: Systematic agricultural mismanagement: the 1985 'bumper' harvest in Zambia. *Journal of Modern African Studies* 24, 257–84.
GOULD, P.R. 1970: Tanzania 1920–63: The spatial impress of the modernization process. *World Politics* 22, 149–70.
GRAHAM, E and FLOERING, I. 1984: *The Modern Plantation in the Third World*. London: Croom Helm.

GREENAWAY, D. and CHONG, H.N. 1988: Industrialization and macroeconomic performance in developing countries under alternative trade strategies. *Kyklos* 41, 419–35.

GRIFFIN, K. 1989: *Alternative Strategies for Economic Development*. London: Macmillan.

GRIMES, O.F. 1976: *Housing For Low-Income Urban Families*. Baltimore MD: Johns Hopkins University Press.

GRINDLE, M. 1986: *State and Countryside: Development Policy and Agrarian Politics in Latin America*. Baltimore MD: Johns Hopkins University Press.

GRINDLE, M. 1988: *Searching for Rural Development: Labour Migration and Employment in Mexico*. Ithaca NY: Cornell University Press.

GWYNNE, R.N. 1985: *Industrialization and Urbanization in Latin America*. London: Croom Helm.

HALL, P. 1984: *The World Cities*. London: Weidenfeld and Nicholson.

HARDIN, G. 1968: The Tragedy of the Commons. *Science* 162, 1243–8.

HARDIN, G. 1977: Ethical implications of carrying capacity. In: Hardin, G. and Baden, J. (eds) *Managing the Commons*. San Francisco CA: W.H. Freeman and Co.

HARDOY, J.E, MITLIN, D. and SATTERTHWAITE, D. 1992: *Environmental Problems in Third World Cities*. London: Earthscan.

HARRIS, J.R. and TODARO, M.P. 1970: Migration, unemployment and development: A two-sector analysis. *American Economic Review* 60, 126–42.

HARRISON, P. 1987: *The Greening of Africa*. London: Paladin.

HENDERSON, J.V. 1988: *Urban Development: Theory, Fact and Illusion*. New York: Oxford University Press.

HIGMAN, B.W. 1976: *Slave Population and Economy in Jamaica 1807–34*. Cambridge: Cambridge University Press.

HILL, H 1982: State enterprises in competitive industry: an Indonesian case study. *World Development* 10, 1015–23.

HILL, H. 1990: Manufacturing industry. In: Booth, A. (ed.) *The Oil Boom and After: Indonesian Economic Policy and Performance in the Soeharto Era*. Singapore: Oxford University Press, 204–57.

HIRSCHMAN, A.O. 1958: *The Strategy of Economic Development*. New Haven CT: Yale University Press.

HIRSCHMAN, A.O. 1977: A generalized linkage approach to development with special reference to staples. *Essays on Economic Development and Cultural Change in Honor of Bert F. Hoselitz*. Chicago, IL: Chicago University Press, 67–98.

HO, S.P.S. 1978: *Economic Development of Taiwan 1860–1970*. New Haven CT: Yale University Press.

HO, S.P.S. 1979: Decontrolled industrialisation and rural development: Evidence from Taiwan. *Economic Development and Cultural Change* 28, 77–96.

HO, S.P.S. 1981: South Korea and Taiwan: development prospects and problems in the 1980s. *Asian Survey*, 21, 1175–96.

HOPPER, D. 1976: The development of agriculture in developing countries. *Scientific American* 235, 3, 196–205.

HOSELITZ, B.F. 1957: Urbanization and economic growth in Asia. *Economic Development and Cultural Change* 6, 42–54.

HUGHES, H. 1988: *Achieving Industrialization in East Asia*. Cambridge: Cambridge University Press.

HUSSAIN, A. and STERN, N. 1991: Economic reform in China. *Economic Policy* 12, April, 141–86.

HYMER, S.H. 1976: *The International Operations of National Firms: A Study of Foreign Direct Investment*. Cambridge MA: MIT Press.

HYUN, Y-S. 1989: *A technology strategy for the Korean motor industry*. Mexico City: IMVP International Policy Forum.

IADB 1992: *Economic and Social Progress in Latin America*. Washington DC: Interamerican Development Bank.

IBGE 1987: IBGE Contrabilidada Social. Rio de Janeiro.

INNES, H. 1920: *The Fur Trade in Canada*. New Haven CT: Yale University Press.

JOHNSON, B.L.C. 1983: *Development in South Asia*. Harmondsworth: Penguin.

JOHNSON, E.A.J. 1970: *The Organization and use of Space in Developing Countries*. Cambridge MA: Harvard University Press.

JOHNSTON, B.F. and MELLOR, J.W. 1961: The role of agriculture in economic development. *American Economic Review* 60, 566–93.

KA, C-M. and SELDON, M. 1986: Original accumulation, equity and late industrialization: the cases of socialist China and capitalist Taiwan. *World Development* 14, 1293–310.

KARMOKOLIAS, Y. 1990: Automotive industry trends and prospects for investment in developing countries. *IFC Discussion Paper* 7, Washington DC: IFC.

KATZMAN, M. 1977: *Cities and Frontiers in Brazil*. Cambridge MA: Harvard University Press.

KELLEY, A.C. and WILLIAMSON, J.G. 1984: *What Drives Third World City Growth?* Princeton NJ: Princeton University Press.

KEMPER, R.V. and FOSTER, G.M. 1975: Urbanization in Mexico: the view from Tzintzuntzan. In: Cornelius, W.A. and Trueblood F. (eds) *Urbanization and Inequality*. London: Sage, 53–75.

KESSEL, N. 1977: Financial aspects of the mining industry in Zambia. In: Newlyn, W.T. (ed.) *The Financing of Economic Development*. Oxford: Clarendon Press.

KEYFITZ, N. 1991: Population growth can prevent the development that would slow population growth. In: Matthews, J.T. (ed.) *Preserving the Global Environment*. New York: W.W. Norton and Co., 39–77.

KILLICK, T. 1983: Kenya, the IMF and the unsuccessful quest for stabilization. In: Williamson, J. (ed.) *IMF Conditionality*. Washington DC: Institute for International Economics, 381–413.

KILLICK, T. 1992: *Explaining Africa's Post-Independence Development Experiences*. ODI Working Paper 60, London: ODI.

KLEINPENNING, J.M.G. 1977: An evaluation of the Brazilian policy for the integration of the Amazon region. *Tijdschrift voor Economische en Sociale Geografie* 68, 297–311.

KOLARS, J.F. and MALIN, H.J. 1970: Population and accessibility: An analysis of Turkish railroads. *Geographical Review* 60, 229–46.

KREUGER, A.O. and TUNCER, B. 1982: An empirical test of the infant industry argument. *American Economic Review* 72, 1142–52.

KRUGMAN, P. 1979: A model of innovation, technology transfer and the world distribution of income. *Journal of Political Economy* 87, 253–6.

KRUGMAN, P. 1986: *Strategic Trade Theory and the New International Economics*. Cambridge MA: MIT.

KRUGMAN, P. 1992: Does the new trade theory require a new trade policy? *World Economy* 15, 423–41.

KRUGMAN, P. 1993: The narrow and broad arguments for free trade. *American Economic Association: Papers and Proceedings*, May, 362–6.

KUZNETS, P.W. 1988: An East Asian model of economic development: Japan, Taiwan and Korea. *Economic Development and Structural Change* 36, 11–43.

KUZNETS, S. 1971: *Economic Growth of Nations*. Cambridge MA: Harvard University Press.

KWASHIMA, T. 1975: Urban agglomeration economies and manufacturing industries. *Papers of the Regional Science Association* 34, 157–75.

LAL, D. 1983: *The Poverty of 'Development Economics'*. London: Institute of Economic Affairs.

LAL, D. 1988: *The Hindu Equilibrium*. Volume 1, Oxford: Clarendon Press.
LAVELL, A. 1972: Industrial development and the regional problem: A case study of Central Mexico. *Regional Studies*, 6, 343–62.
LEEMING, F. and POWELL, S. 1990: Rural China: old problems and new solutions. In: Cannon, T. and Jenkins, A. (eds) *The Geography of Contemporary China*. London: Routledge, 133–67.
LEVY, M.B. and WADYCKI, W.J. 1974: What is the opportunity cost of moving? Reconsideration of the effects of distance on migration. *Economic Development and Cultural Change* 22, 198–214.
LEWIS, S.R. 1982: Development problems of the mineral-rich countries. *Williams College Centre for Development Economics Research Memo* 74, Williamstown MA: Williams College.
LEWIS, W.A. 1954: Economic development with unlimited supplies of labour. *Manchester School of Economic and Social Studies* 20, 139–92.
LEWIS, W.A. 1977: The evolution of the international economic order. Discussion Paper 74, Princeton University NJ: Woodrow Wilson Research Program in Development Studies.
LEWIS, W.A. 1978: *Growth and Fluctuations*. London: George Allen and Unwin.
LI, W-Y. 1990: Contemporary spatial issues. In: Linge, G.J.R. and Forbes, D.K. (eds) *China's Spatial Economy*. Hong Kong: Oxford University Press, 59–84.
LIANG, K. and LIANG, C.H. 1988: Development policy formulation and future priorities in the Republic of China. *Economic Development and Cultural Change* 36, 67–101.
LIPTON, M. 1977: *Why Poor People Stay Poor*. London: Temple Smith.
LOSER, C. and KALTER, E. 1992: Mexico; the strategy to achieve economic growth. *Occasional Paper* 99, Washington DC: IMF.
LUCAS, R.E.B. 1988: India's industrial policy. In: Lucas, R.E.B. and Papenak, G.F. (eds) *The Indian Economy: Recent Developments and Future Prospects*. London: Westview Press, 185–202.
LUCKE, M. 1988: Economies of scale in protected manufacturing in the developing countries; Case of the Brazilian passenger car industry. *Industry and Development* 24, 35–55.
LUSTIG, N. 1993: *The Mexican Economy*. Washington DC: Brookings Institution.
MACFARQUHAR, E. 1987: India, Push Comes to Shove. *The Economist*, May 9, centre supplement.
MAHENDRA DEV, S., SURYANARAYANA, M.H and PARIKH, K.S. 1992: Rural poverty in India: Incidence, issues and policies. *Asian Development Review* 10, 35–66.
MAHON, J.E. 1992: Was Latin America too rich to prosper? Structural and political obstacles to export-led growth. *Journal of Development Studies* 28, 241–63.
MAKHIJANI, A. 1975: *Energy and Agriculture in the Third World*. Cambridge MA: Ballinger Publishing Company.
MATHER, D.J. 1989: Deforestation in Brazil's Amazon region: Magnitude, rates and causes. In: Schramm, G. and Warford, J.J. (eds) *Environmental Management and Economic Development*. Bethesda MD: Adler and Adler, 191–234.
MELLOR, J.W. 1976: *The New Economics of Growth: A Strategy for India and the Developing World*. Ithaca NY: Cornell University Press.
MERA, K. 1973: On the urban agglomeration and economic efficiency. *Economic Development and Cultural Change* 21, 309–24.
MESAROVIC, M. and PESTEL, E. 1974: *Mankind at the Turning Point*. New York: Dutton.
MEYER, J.R., KAIN, J.F. and WOHL, M. 1965: *The Urban Transportation Problem*. Cambridge MA: Harvard University Press.
MIKESELL, R.F. 1992: *Economic Development and The Environment*. London: Mansell.

MORISHIMA, M. 1982: *Why has Japan Succeeded? Western Technology and Japanese Ethos*. Cambridge: Cambridge University Press.

MOSCOVITCH, E. 1969: Evaluating the allocation of resources to urban development. In: Rodwin, L. (ed.) *Planning Urban Growth and Regional Development: The Experience of the Guayana Program of Venezuela*. Cambridge MA: MIT Press, 378–99.

MUNDLE, S. 1990: Food, finance and foreign trade; the limits of high growth in India. Working Paper, Cambridge University.

MUNRO, J.M. 1974: Migration in Turkey. *Economic Development and Cultural Change* 22, 634–53.

MURPHY, K.M. *et al.* 1989: Industrialization and the big push. *Journal of Political Economy* 97: 1003–26.

MYINT, H. 1964: *The Economics of the Developing Countries*. London: Hutchinson.

MYRDAL, G. 1957: *Economic Theory and Underdeveloped Regions*. London: Methuen.

NANKANI, G. 1979: Development problems of mineral exporting countries. *World Bank Staff Working Paper* 354, Washington, DC: World Bank.

NAYYAR, D. 1988: India's export performance 1970–85: Underlying factors and constraints. In: Lucas, R.E.B. and Papenak, G. (eds) *The Indian Economy*. Boulder CO: Westview Press, 217–52.

NEWCOMBE, K. 1989: An economic justification for rural afforestation: The case of Ethiopia. In: Schramm, G. and Waterford, J.J. (eds) *Environmental Management and Economic Development*. Baltimore MA: Johns Hopkins University Press, 117–38.

NORTH, D. C. 1955: Location theory and regional economic growth. *Journal of Political Economy* 63, 248–58.

OHNO, K. and IMAOKA, H. 1987: The experience of dual industrial growth: Taiwan and Korea. *The Developing Economies* 35, 310–24.

O'RIORDAN, T. 1988: The politics of sustainability. In: Turner, R.K. (ed.) *Sustainable Environmental Management: Principles and Practice*. London: Belhaven Press.

OWEN, W. 1968: *Distance and Development: Transport and Communications in India*. Washington DC: Brookings Institution.

OWEN, W. 1987: *Transportation and World Development*. London: Hutchinson.

PARK, Y.C. 1986: Foreign debt, balance of payments and growth prospects: the case of the republic of Korea 1965–88. *World Development* 14, 1019–58.

PEARCE, D., BARBIER, E. and MARKANDYA, A. 1990: *Sustainable Development*. London: Earthscan.

PEATTIE, L. 1975: 'Tertiarization' and urban poverty in Latin America. In: Cornelius, W.A. and Trueblood, F.M. (eds) *Urbanization and Inequality: The Political Economy of Urban and Rural Development in Latin America*. London: Sage, 109–23.

PERKINS, D.H. 1988: Reforming China's economic system. *Journal of Economic Literature* 26, 601–45.

PERKINS, D. and SYRQUIN, M. 1989: Large countries: The influence of size. In: Chenery, H. and Srinivasan, T.M. (eds) *Handbook of Development Economics*. Volume 2, Amsterdam: North Holland, 1691–753.

PERKINS, D.H. and YUSUF, S. 1984: *Rural Development in China*. Oxford: Oxford University Press.

PESKIN, H.M. 1993: Sustainable resource accounting. Noordwijk, the Netherlands: Paper presented to the International Expert Meeting on Strategies for Sustainable Resource Management and Resource Use, June 3–4.

PIMENTAL, D. *et al.* 1973: Food production and the energy crisis. *Science* 182, 443–9.

PINGALI, P. and BINSWANGER, H. 1988: Population density and farming systems. In: Lee, R.E., Arthur, W.B., Kelley, A.C., Rodgers, G. and Srinivasan, T.N. (eds) *Population, Food, and Rural Environment*. Oxford: Oxford University Press, 51–76.

PINTO, B. 1987: Nigeria during and after the oil boom; a policy comparison with Indonesia. *World Bank Economic Review* 1, 419–25.

POLLARD, N. 1981: The Gezira Scheme: A study in failure. *Ecologist* 11, 31.
PORTEOUS, J.D. 1973: The corporate state: copper production in Chile. *Canadian Geographer* 17, 113–26.
POSCO 1988: *Posco's Present and Future*. Seoul: Posco.
PREBISCH, R. 1950: The economic development of Latin America and its principal problems, ECLA/UN. Also published in: *Economic Bulletin for Latin America* 7, 1962, 1–22.
PREBISCH, R. 1963: *Toward a Dynamic Development Policy for Latin America*. New York: UN.
PRYBYLA, J.S. 1987: *Market and Plan Under Socialism*. Stanford CA: Hoover Institution Press.
QUIZON, J. and BINSWANGER, H. 1986: Modelling the impact of agricultural growth and government policy on income distribution in India. *World Bank Research Observer* 1
RADETZKI, M. 1985: *State Mineral Enterprises*. Washington DC: RFF.
RANIS, G. and MAHMOOD, S. 1992: *The Political Economy of Development Policy Change*. Oxford: Basil Blackwell.
RAO, T.R. 1984: Scenarios for the Indian iron and steel industry. *Long Range Planning* 17, no. 4, 91–101.
RAPPAPORT, R.A. 1971: The flow of energy in an agricultural society. *Scientific American (Energy and Power)*, 68–80.
RAWSKI, T.G. 1979: *Economic growth and Employment in China*. New York: Oxford University Press.
REPETTO, R. 1989: Economic incentives for sustainable production. In: Schramm, G. and Warford, J. (eds) *Environmental Management and Economic Development*. Baltimore MD: Johns Hopkins University Press, 69–86.
REPETTO, R. 1992: Accounting for environmental assets. *Scientific American* 266, 6, 64–70.
REPETTO, R., WELLS, M., BEER, C. and ROSSINI, F. 1987: *Natural Resource Accounting for Indonesia*. Washington D.C.: World Resources Institute.
REVELLE, R. 1984: The world supply of agricultural land. In: Simon, J.L. and Kahn H. (eds) *The Resourceful Earth*. Oxford: Basil Blackwell, 184–201.
REYNOLDS, C.W. 1970: *The Mexican Economy: Twentieth Century Structure and Growth*. New Haven CT: Yale University Press.
RICHARDSON, H.W. 1973: *The Economics of Urban Size*. London: Saxon House.
RICHARDSON, H. 1993: Efficiency and welfare in LDC mega-cities. In: Kasarda, J.D. and Parnell, A.M. (eds) *Third World Cities*. London: Sage Publications, 32–57.
RICHARDSON, H.W. and RICHARDSON, M. 1975: The relevance of growth centre strategies to Latin America. *Economic Geography* 51, 163–78.
RISKIN, C. 1988: *China's Political Economy; The Quest For Development Since 1949*. Oxford: Oxford University Press.
ROBINSON, R.J. and SCHMITZ, L. 1989: Jamaica: Navigating through a troubled decade. *Finance and Development* 26, no. 4, 30–3.
ROEMER, M. 1985: Dutch disease in developing countries: Swallowing the bitter economic medicine. In: Lundahl, M. (ed.) *The Primary Sector in Economic Development*. London: Croom Helm, 234–52.
RONDINELLI, D.A. 1991: Asian urban development policies in the 1990s: from growth control to urban diffusion. *World Development* 19, 791–803.
ROSENSTEIN-RODAN, P.N. 1943: Problems of industrialization of eastern and southern Europe. *The Economic Journal* 53, 202–11.
ROSTOW, W.W. 1990: *The Stages of Economic Growth: A Non-Communist Manifesto*, third edition. Cambridge: Cambridge University Press.
ROTHENBERG, J. 1987: Space, interregional economic relations and structural reform in China. *International Regional Science Review* 11, 5–22.

ROY, K.C., TIDSELL, C. and ALAUDDIN, M. 1992: Rural–urban migration and poverty in South Asia. *Journal of Contemporary Asia* 22, 57–72.

RUMELT, R.P. 1986: *Strategy, Structure and Economic Performance*. Boston MA: Harvard Business School Press.

SACHS, J.D. 1985: External debt and macroeconomic performance in Latin America and East Asia. *Brookings Papers on Economic Activity*, no. 2, 523–73.

SACHS, J.D. 1989: Social conflict and populist policies in Latin America. *NBER Working Paper*, 2897. Mimeo, Cambridge MA: National Bureau of Economic Research.

SAHA, S.K 1990: Industrialization and interregional disparities in post-colonial India: Towards a new regional policy. *Tijdschrift voor Economische en Sociale Geografie* 81, 93–109.

SALLEH, I.M. and MEYANATHAN, S.D. 1993: *Malaysia: Growth, Equity and Structural Transformation*. Washington DC: World Bank.

SANDBROOK, R. 1986: The state and economic stagnation in tropical Africa. *World Development* 14, 319–32.

SANDESARA, J.C. 1986: Industrial production and employment in the seventh plan – Two quick comments. *The Indian Economic Journal* 33, 92–7.

SAVAGE, C.I. 1966: *An Economic History of Transport*. London: Hutchinson.

SCHORSCH, L.L. 1989: Minerals trade and commercial policy: The case of steel. *Resources Policy* 15, 169–87.

SCHULTZ, T.W. 1964: *Transforming Traditional Agriculture*. New Haven CT: Yale University Press.

SCOTT, I. 1982: *Urban and Spatial Development in Mexico*. Baltimore MD: Johns Hopkins University Press.

SEERS, D. 1981: *Dependency Theory: A Critical Reassessment*. London: Francis Pinter.

SEGAL, D. 1976: Are there returns to scale in city size? *Review of Economics and Statistics* 58, 339–50.

SELOWSKY, M. and van der TAK, H.G. 1986: The debt problem and growth. *World Development* 14, 1107–24.

SHAFER, M. 1983: Capturing mineral multinationals: advantage or disadvantage? *International Organization* 22, 93–119.

SHARPLEY, J. 1983: Economic management and IMF conditionality in Jamaica. In: Williamson, J. (ed.) *IMF Conditionality*. Washington DC: Institute for International Economics, 233–62.

SHEPHERD, G. 1991: The communal management of forests in the semi-arid and sub-humid regions of Africa: past practice and prospects for the future. *Development Policy Review* 9, 151–76.

SHIRLEY, M. 1983: Managing state-owned enterprises. *World Bank Staff Working Paper* 577, Washington DC: World Bank.

SIMONSEN, M.H. 1989: Brazil. In: Dornbusch, R. and Helmers, F.L.C. (eds) *The Open Economy*. New York: Oxford University Press, 285–306.

SINGER, H.W. 1950: The distribution of gains between investing and borrowing countries. *American Economic Review: Papers and Proceedings* 40, 473–5.

SKELDON, R. 1990: *Population Mobility in Developing Countries*. London: Belhaven Press.

SMITH, N.J.H. 1981: Colonization lessons from a tropical forest. *Science* 214, 755–61.

SONG, B-N. 1990: *The Rise of the Korean Economy*. Hong Kong: Oxford University Press.

SOON, T-W. and TAN, C.S. 1993: *Singapore: Public Policy and Economic Development*. Washington DC: World Bank.

SPEARE, A. 1974: Urbanization and migration in Taiwan. *Economic Development and Cultural Change* 22, 302–19.

SQUIRE, L. 1993: Fighting poverty, *American Economic Association: Papers and Proceedings* 377–82.

SRINIVASAMURTHY, A.P. 1981: *Investment Allocation in Indian Planning*. Bombay: Himalaya Publishing House.

STAUFFER, T. 1985: Accounting for wasting assets: Measurements of income and dependency in oil-rentier states. *Journal of Energy and Development* 11, 69–93.

STERN, J.J. 1990: Industrial targeting in Korea. *HIID Development Discussion Paper* 343, Cambridge MA: HIID.

STEWART, D. and GHANI, E. 1992: Externalities, development and trade. In: Helleiner, G. (ed.) *Trade Policy, Industrialization and Development*. Oxford: Clarendon Press, 122–54.

STILLMAN, B. 1973: Selectivity in rural–urban migration; The case of Huaylas, Peru. In: Southall, A. (ed.) *Urban Anthropology; Cross Cultural Studies of Urbanization*. Oxford: Oxford University Press, 351–72.

STOBAUGH, R. 1970: The neo-technology account of international trade; The case of petrochemicals. *Journal of International Business Studies* 2, no. 2, 41–60.

STORPER, M. 1991: *Industrialization, Economic Development and the Regional Question in the Third World*. London: Pion.

SUH, S-M. 1992: The economy in historical perspective. In: Corbo, V. and Suh, S-M. (eds) *Structural Adjustment in Newly Industrialised Countries: The Korean Experience*. Baltimore MD: Johns Hopkins University Press.

SUMMERS, L.H. 1992: Summers on sustainable growth. *The Economist*, May 30, p. 91.

SUMMERS. L.H. and EASTERLY, W.R. 1992: Culture is not to blame. *Financial Times*, April 15.

SVIEKAUSKAS, L. 1975: The productivity of cities. *Quarterly Journal of Economics* 89, 393–413.

SYRQUIN, M. and CHENERY, H.B. 1989: Patterns of Development 1950 to 1980. *World Bank Discussion Paper* 41, Washington DC: World Bank.

TAAFFE, E., MORRILL, R. and GOULD, P. 1963: Transport evolution in underdeveloped countries: A comparative analysis. *Geographical Review* 53, 503–29.

THOBURN, J.T. 1977: *Primary Commodity Exports and Economic Development*. London: Wiley.

THOMAS, D.G. 1993: Sandstorm in a teacup? Understanding desertification. *Geographical Journal* 159, 318–31.

TICHY, G. 1989: The product-cycle revisited: Some extensions and clarifications. *Research Memorandum* 8901, University of Graz: Department of Economics.

TIETENBERG, T. 1992: *Environmental and Natural Resource Economics*. New York: HarperCollins.

TIFFIN, M., MORTIMORE, M. and GICHUKI, F. 1994: *More People, Less Erosion*. Chichester: Wiley.

TOLBA, M.K. *et al.* 1992: *The World Environment 1972–92*. London: Chapman & Hall.

TONGEREN, J.VAN, LUTZ, E., GOMEZ LUNA, M. and MARTIN, F.G. 1991: Integrated Environmental and Economic Accounting: A Case Study for Mexico. *Environment Working Paper* 50, Washington DC: World Bank.

TOULMIN, C., SCOONES, I. and BISHOP, J. 1992: The future of Africa's drylands: Is local resource management the answer? In: Homberg, J. (ed.) *Policies for a Small Plant*. London: Earthscan, 225–57.

TSIANG, S.C. 1984: Taiwan's economic miracle: lessons in economic development. In: Harberger, A.C. (ed.) *World Economic Growth*. San Francisco CA: ICS Press, 301–26.

TURNER, L. and MCMULLEN, N. 1982: *The Newly Industrializing Countries: Trade and Adjustment*. London: George Allen and Unwin.

TURNER, R.K., PEARCE, D. and BATEMAN, I. 1994: *Environmental Economics*. London: Harvester Wheatsheaf.

UNIDO 1985: *India*. Regional and Country Studies Branch, Vienna: Unido.

UNITED NATIONS 1950–85: *UN Statistical Yearbook*. New York: UN.

UNITED NATIONS 1992: *World Investment Report*. New York: UN.

VERMEER, D.E. 1970: Population pressure and crop rotational changes among the Tiv of Nigeria. *Annals of the Association of American Geographers* 60, 299–314.

VERNON, R. 1966: International investment and international trade in the product cycle. *Quarterly Journal of Economics* 80, 190–207.

VERNON, R. 1979: The product cycle hypothesis in a new international environment. *Oxford Bulletin of Economics and Statistics* 41, 255–67.

WADE, R. 1990: *Governing the Market: Economic Theory and the Role of Government in East Asian Industrialization*. Princeton NJ: Princeton University Press.

WALTON, J. 1975: Internal colonialism: Problems of definition and measurement. In: Cornelius, W.A. and Trueblood, F.M. (eds) *Urbanization and Inequality*. London: Sage, 29–50.

WALTON, J. 1978: Guadalajara: Creating the divided city. In: Cornelius, W.A. and Kemper, R.V. (eds) *Metropolitan Latin America: The Challenge and Response*. London: Sage, 25–50.

WANGING GUO 1988: The transformation of Chinese regional policy. *Development Policy Review* 6, 29–50.

WARD, B. 1976: *The Home of Man*. New York: W.W. Norton.

WARD, B. and DUBOS, R. 1972: *Only One Earth*. Harmondsworth: Penguin.

WARD, P.M. 1976: The squatter settlement as slum or housing solution. *Land Economics* 52, 330–46.

WARD, P.M. 1990: *Mexico City*. London: Belhaven Press.

WATKINS, M. 1963: A staple theory of economic growth. *The Canadian Journal of Economics and Political Science* 29, 141–58.

WAUGH, D. 1987: *The World*. London: Nelson.

WHEELER, D. 1984: Sources of stagnation in Sub-Saharan Africa. *World Development* 12, 1–23.

WILLBANKS, T.J 1972: Accessibility and technological change in India. *Annals of the Association of American Geographers* 62, 427–36.

WILLIAMS, M.L. 1975: The extent and significance of the nationalization of foreign-owned assets in the developing countries 1956–72. *Oxford Economic Papers* 27, 260–73.

WILLIAMSON, J.G. 1965: Regional inequality and the process of national development: a description of the patterns. *Economic Development and Cultural Change* 13, 3–45.

WILLIAMSON, J. 1993: Democracy and the 'Washington Consensus'. *World Development* 21, 1329–36.

WILLIAMSON, J. and MILNER, C. 1991: *The World Economy*. London: Harvester Wheatsheaf.

WOMACK, J.P., JONES, D.T. and ROOS, D. 1990: *The Machine That Changed the World*. New York: Rawson Associates.

WOOD, A. 1986: Growth and structural change in large low-income countries. *World Bank Staff Working Paper* 763, Washington DC: World Bank.

WOOD, A. 1988: Global trends in real exchange rates 1960 to 1984. *Discussion Paper* 35, Washington DC: World Bank.

WOOD, A. 1990: China's economic system. *Development Economics Research Programme* CP 12. London: LSE.

WORLD BANK 1981: *China: Socialist Economic Development*. Report 3391-CHA. Washington DC: World Bank.

WORLD BANK 1982: *Brazil: Industrial Policies and Manufactured Exports*. Washington DC: World Bank.

WORLD BANK 1985a: *China: Economic Structure in International Perspective*. Annex 5 to *China: Long-Term Development Issues and Options*. Washington DC: World Bank.

WORLD BANK 1985b: *China: Long-Term Development Issues and Options*. Baltimore: Johns Hopkins University Press.

WORLD BANK 1987: *Korea: Managing the Industrial Transition*. Washington DC: World Bank.

WORLD BANK 1988: *Papua New Guinea: Policies and Prospects for Sustained and Broad-Based Growth*. Washington DC: World Bank.

WORLD BANK 1989a: *Sub-Saharan Africa: From Crisis to Sustainable Growth*. Washington DC: World Bank.

WORLD BANK 1989b: *India: An Industrializing Economy in Transition*. Washington DC: World Bank.

WORLD BANK, 1989c: *China: Macroeconomic Stability and Economic Growth*. Report 7483-CHA. Washington DC: World Bank.

WORLD BANK, 1990a: *World Development Report 1990*. New York: Oxford University Press.

WORLD BANK, 1990b: *China: Between Plan and Market*. Washington DC: World Bank.

WORLD BANK 1990c: *World Tables 1990*. Washington DC: World Bank.

WORLD BANK 1991: *World Tables 1991*. Washington DC: World Bank.

WORLD BANK, 1992a: *World Development Report 1992*. New York: Oxford University Press.

WORLD BANK 1992b: Strategy For African Mining. *World Bank Technical Paper* 181. Washington DC: World Bank.

WORLD BANK, 1993a: *The East Asian Miracle*. New York: Oxford University Press.

WORLD BANK 1993b: *World Development Report 1993*. Washington DC: World Bank.

WORLD COMMISSION ON ENVIRONMENT AND DEVELOPMENT 1987: *Our Common Future*. Oxford: Oxford University Press.

WORLD RESOURCES INSTITUTE 1992: *World Resources 1992–93*. Oxford: Oxford University Press.

WU, Q. 1988: South Korea and Taiwan: A comparative analysis of economic development. *IDS Discussion Paper* 252, Brighton: IDS, University of Sussex.

YOO, J-H 1990: The industrial policy of the 1970s and the evolution of the manufacturing sector in Korea. *KDI Working Paper* 9017. Seoul: KDI.

YOUNG, S. and YOO, G.H. 1982: The basic role of industrial policy and the reform proposal for the protection regime in Korea. Mimeo, Seoul: Korean Development Institute.

ZWEIG, D. 1987: From village to city: Reforming urban–rural relations in China. *International Regional Science Review* 11, 43–58.

Glossary

Absolute poverty Where incomes are barely sufficient to meet basic subsistence needs of food, shelter, clothing and health care.
Absorptive capacity The ability of a country to absorb foreign capital effectively.
Autarky A relatively closed (i.e. self-sufficient) economy or productive sector.

Balance of payments The statistic describing a country's financial transactions with the outside world.
Big push A development strategy which uses a massive investment in complementary industry and infrastructure in order to capture the potential internal and external economies of scale.

Capital-intensive output Production which uses a high proportion of capital relative to other factors of production such as land and labour, per unit of output.
Cartel A grouping of producers agreeing to limit their output in order to achieve higher prices and returns.
Comparative advantage Where one country has an advantage over another in producing a commodity because it can do so at lower opportunity cost in terms of the foregone alternative production opportunities.
Cost-benefit analysis A method for comparing the actual and potential private and social costs of an economic decision with the private and social benefits. Projects which yield the highest ratio of benefits to costs are preferred.
Current account The section of the balance of payments which shows the value of a country's visible (e.g. commodity trade) and invisible (e.g. tourism services) exports and imports with the rest of the world.

Debt service The sum of the interest payments and repayments of principal on external public debt and publicly guaranteed debt.
Debt-service ratio The ratio of annual interest and principal repayments to export receipts for the year.
Demonstration effects The impact of foreign lifestyles (including consumption habits) on a country's nationals.
Dependence A situation in which developing countries are believed to rely on a developed country's policies in order to experience their own economic growth.
Dependency ratio The ratio of the population considered to be economically inactive (usually aged 0–15 and above 65 as well as nursing mothers) to the working population.
Devaluation The reduction in a country's official exchange rate relative to all other countries.
Diminishing returns The decline in marginal productivity with expanding output.
Doubling rate The period over which a given population doubles in size. It may be approximated by dividing the annual growth rate into 70.
Dutch Disease The contraction of output in the tradeables sector (farming and manufacturing) due to the strengthening of the exchange rate associated with a commodity boom.

Economic efficiency Producing the maximum output possible for a given level of market demand with cost-minimizing techniques of production.

Economic enclave Economic activity which provides minimal economic stimulus to the local area and a maximum stimulus to distant regions.
Economies of scale The economies resulting from the expansion of production by a firm or sector, leading to higher output at lower unit cost. Internal economies accrue to the individual firm whereas external economies accrue to a group of firms as a result of the productive advantages arising from the scale of their combined production.
Effective rate of protection The degree of protection on value-added as opposed to the final price of the product, including all purchased inputs used in the production process.
Energy/materials intensity of GDP Ratio of total energy/materials consumption per unit of GDP (usually expressed per $000 of per capita income).
Entry barriers Restrictions on the start-up of new firms such as licences, specialized technology and very large capital requirements.
Environmental accounting A set of satelite accounts to the standard national economic accounts which assign monetary values to net changes in the total stock of environmental resources, including the measurement of environmental degradation.
Environmental capital A measure of the stock of natural resources such as minerals and forests.
Exchange rate Rate at which central banks exchange one country's currency for another.

Factors of production The resources required to produce a good or service (usually comprizing capital, land and labour).
Financial liberalization The reduction of state intervention in financial markets to allow market forces to play a greater role in setting, for example, interest rates.
Flexible exchange rate A national exchange rate which is free to move in response to changes in demand and supply for the currency, as driven by international demand.
Flow resources Those natural resources in which annual consumption is equal to or less than the annual rate of their formation. Such resources are also termed 'renewable' resources.
Foreign aid The international transfer of loans or grants from one country to another, either directly (bilateral aid) or indirectly through intermediary agencies (multilateral aid).
Free trade Trade which occurs without any barriers such as tariffs, quotas or other restrictions.
Fund resources Those natural resources where the annual rate of consumption exceeds the annual rate of formation. Usually applies to minerals but can also apply to mismanaged flow (renewable) resources such as over-exploited forests.

Gini coefficient A measure of income inequality ranging from 0 (perfect equality) to 1 (perfect inequality) which is depicted graphically by dividing the area between the 45 degree 'perfect equality line' and the actual line traced by the cumulative increase in population and income share.
Gross domestic investment The expenditures on additional fixed assets by both the public and private sectors and including any net change in inventories.
Gross domestic product (GDP) The final output of goods and services produced by a country's economy, within the nation's territory, by both residents and non-residents, regardless of its allocation between domestic and foreign claims.
Gross national product (GNP) The total domestic and foreign output attributable to residents (i.e. nationals) of a country. It comprises GDP plus factor incomes accruing to residents from abroad, minus the income earned in the domestic economy by persons abroad.

Growth poles Regions, usually urban centres, which are more advanced than those around them and which benefit from the capture of the external economies of scale.

Human capital Productive investments embodied in people such as education.

Imperfect competition A market in which producers can exercise some degree of control over prices, usually because there are relatively few producers so that the assumption made in perfect competition (of many producers) does not hold.

Import substitution (ISI) State intervention through tariffs, quotas and incentives to encourage domestic manufacturers to set up factories to replace hitherto imported items.

Increasing returns A more than proportionate increase in output with increasing scale of production (i.e. the economies of scale).

Incremental capital output ratio (ICOR) The amount of capital required to raise output by one unit.

Infant industry A newly-established industry, usually enjoying the protection of tariff barriers and other inducements to invest.

Inflation The phenomenon of rising prices, usually inferring to some above-normal price trend.

Informal sector The section of the urban economy characterized by large numbers of small individually-owned or family firms operating under intense competition in petty retail trade, services and craft industries.

Infrastructure The physical and financial investments made in transportation networks, utilities and public services such as health and education.

Inputs Goods and services, such as hours of work or raw materials, used in the production process.

Intermediate goods Goods which are used as inputs into further levels of production such as iron ore in steel manufacture or steel manufacture in the case of auto assembly.

Investment The part of national income (expenditure) devoted to the production of capital goods over a given period of time. A useful distinction is made between gross investment which is the total expenditure in new capital goods and net investment which is the additional capital goods produced after deducting depreciation (wear and tear).

Inward-orientation Policies which stress autarky or self–reliance and often associated with import substitution.

Labour productivity The level of output per unit of labour input, usually expressed in terms of man-hours per unit output.

Macroeconomics The approach within economics which examines the relationships between the broad economic aggregates such as national income, saving, consumption and investment.

Marginal cost The additional total cost to a producer as a result of changing output by one further unit.

Market failure A situation in which the existence of market imperfections (such as monopoly power, labour market inflexibility) weakens the functioning of a free-market economy so that it fails to realize the theoretical benefits of such an economy. Market failure occurs frequently, for example in relation to environmental objectives, and is often used to justify state intervention.

Mineral stabilization fund A fund established to capture a fraction of the rents (excess returns) accruing from a commodity boom in order to dampen the disruptive effects of the commodity boom/bust cycle.

Monopoly A market in which a single producer provides a product with no close substitute.

Multinational corporation An international firm with headquarters in one country but branches in a wide range of countries.

National income The total monetary value of all final goods and services produced in an economy over a period of time.
Neoclassical economics The economics of the capitalist market system characterized by perfect competition, private enterprise, profit maximization and consumer sovereignty. It emphasizes the efficient allocation of scarce resources among competing uses through the deployment of the price mechanism to signal the relative scarcity of goods and services.
Net present value The value of a future stream of benefits discounted to the present using an appropriate discount rate (which usually reflects the opportunity cost [i.e. investment opportunities foregone] of alternative investment options).
Nominal rate of protection The tariff levied on the total cost of imported goods (compare net effective protection).

Oligopoly Markets in which there are few sellers and many buyers of similar but differentiated products.
Open economy An economy which encourages free trade. Sometimes described as an outward-oriented economy.
Opportunity cost The real value of the resources used in the most desirable alternative production opportunity. This concept is used to reflect the benefits foregone by deciding upon a particular investment decision.
Outward-oriented policies Policies which encourage free trade and the free movement of capital and workers.
Overvalued exchange rate An official exchange rate which is set at a higher level than the market value of the currency. Such a policy cheapens the real cost of imports while raising the real cost of exports.

Per capita income The total GNP of the country divided by the total population.
Perfect competition A market characterized by many buyers and sellers of homogeneous products or services with perfect knowledge and free entry so that no individual buyer or seller can influence the price.
Physical capital Tangible man-made investment goods (sometimes known as 'produced capital' to distinguish it from environmental capital and human capital).
Political economy The merging of economic analysis with practical politics so that economic prescriptions are viewed within their political context.
Pollution tax A tax levied on the quantity of pollutants emitted into the environment, which is usually intended to provide an incentive for abatement.
Prebisch–Singer thesis The argument that primary product export–led growth results in a decline in the developing countries' terms of trade.
Primary import substitution The first stage of the East Asian development model in which import substitution is initially established.
Primary products Products derived from all extractive operations: forestry, farming, mining and fishing.
Privatization The sale of state assets to private companies.
Product cycle A model which suggests that the sales of a product exhibit an underlying regularity which traces an S-shaped curve over time in which a pioneering phase of relatively slow growth is followed by a youthful phase of rapid expansion before maturation of demand causes sales to decelerate.
Production function The technological or engineering relationship between the quantity of output produced and the quantity of inputs required to produce it.
Public sector The section of the economy whose activities are owned and managed by the state.
Purchasing power parity A measure of how much can be purchased with a given currency, taking account of variations in the real prices of goods and services that occurs within different economies. For example, many services tend to be much cheaper in low–income countries compared with high-income countries.

Quintile One-fifth (20 per cent) of any numerical quantity.
Quota A physical limit on the quantity of a product that can be imported into a country.

Redistributive policy A policy designed to transfer resources from rich to poor in order to reduce income inequality.
Regressive tax A tax structure in which the fraction of income paid in taxes decreases as incomes rise.
Relative poverty A condition in which, although individuals may have escaped absolute poverty, their sense of deprivation persists due to the demonstration effect of the lifestyles of richer groups in society.
Rent In macroeconomics this is the share of national income going to the owners of the productive assets, whereas in microeconomics it represents a return in excess of that needed for an efficient producer to remain in production.
Rent-seeking The effort expended by economic agents in seeking to benefit from the rents (excessive returns) created by price distortions and physical controls arising from excessive state intervention, such as infant industry support.

Scale–sensitive Those sectors characterized by the existence of the internal economies of scale.
Secondary economic activity The manufacturing sector of the economy which uses raw materials and intermediate goods to produce final goods or additional intermediate products.
Shifting cultivation A farming system which rotates the fields over a number of years. Sometimes known as 'swidden', 'milpa' or 'slash and burn'.
Skewed income distribution An income distribution which diverges from an equitable pattern.
Social discount rate The rate at which society discounts potential future social benefits. The rate is usually lower than the private discount rate, reflecting the greater ability of societies (as opposed to individuals or firms) to bear risk.
Stabilization policy A package of fiscal and monetary policies designed to curb inflation, reduce budget deficits and improve the balance of payments.
State-owned enterprises Public firms and parastatal organizations (such as crop marketing boards) which are owned and managed by the state.
Structural adjustment loans Loans provided by the World Bank which require conditions to be met which reduce state intervention and restructure the economy away from unsustainable levels of domestic consumption.
Structural change The transformation of the production structure of the economy from predominantly subsistence agriculture to a modern service economy.
Subsistence Production which is mainly for personal consumption, usually involving little trade but relatively high levels of risk and uncertainty.
Sustainable development Development which permits future generations to enjoy living standards at least as high as those enjoyed by the present generation.

Tariff A fixed percentage tax on imported goods.
Technological maturation The achievement of international levels of technological competence, including the ability of a firm to make in-house changes to production technology.
Terms of trade The ratio of the average price of a country's exports to the average price of its imports. The terms of trade deteriorate when this ratio declines, i.e. when the prices of imports rise compared with the rise of exports.
Total factor productivity The total monetary value of all units of output per unit of each and every factor of production in the economy.
Trade liberalization The removal of barriers to free trade such as tariffs and quotas.

Transfer pricing An accounting procedure used by MNCs in which vertical integration between production stages allows goods to be transferred within the corporation at prices set to minimize taxation. For example, in high tax countries a firm may set prices equal to costs, thereby declaring no profits and evading taxes while securing a relatively cheap input into the next stage of the production chain where profits will be greater and the tax rate may be less.

Transfers The payment of subsidies such as welfare payments within an economy.

Trickle-down effects The notion that advances in economic activity will ripple through an economy to benefit all participants within it.

Underemployment A condition in which individuals work less than they would like.

Value-added Comprizes the returns to capital and labour and is defined as the price of the good or service minus the cost of purchased inputs.

Vested interest groups Collections of individuals who, having secured advantages through specific government policies, seek to perpetuate such policies by blocking reform.

Vicious cycle The perpetuation of an undesirable socio-economic condition through a self-reinforcing chain of events.

Index